HOLT SCIENCE & TECHNOLOGY

Cells, Heredity and Classification

Interactive Textbook

HOLT, RINEHART AND WINSTON
A Harcourt Education Company
Orlando • Austin • New York • San Diego • London

Printed in the United States of America

ISBN 0-03-095814-8

1 2 3 4 5 6 7 082 10 09 08 07 06

Contents

CHAPTER 6 The History of Life on Earth

CHAPTER 7 Classification

Name ______________________ Class ______________ Date ______________

CHAPTER 1 Cells: The Basic Units of Life

SECTION 1

The Diversity of Cells

National Science Education Standards

LS 1a, 1b, 1c, 2c, 3b, 5a

BEFORE YOU READ

After you read this section, you should be able to answer these questions:

- What is a cell?
- What do all cells have in common?
- What are the two kinds of cells?

What Is a Cell?

Most cells are so small that they cannot be seen by the naked eye. So how did scientists find cells? By accident! The first person to see cells wasn't even looking for them.

A **cell** is the smallest unit that can perform all the functions necessary for life. All living things are made of cells. Some living things are made of only one cell. Others are made of millions of cells.

Robert Hooke was the first person to describe cells. In 1665, he built a microscope to look at tiny objects. One day he looked at a piece of cork. Cork is found in the bark of cork trees. Hooke thought the cork looked like it was made of little boxes. He named these boxes *cells*, which means "little rooms" in Latin.

STUDY TIP

Organize As you read this section, make lists of things that are found in prokaryotic cells, things that are found in eukaryotic cells, and things that are found in both kinds of cells.

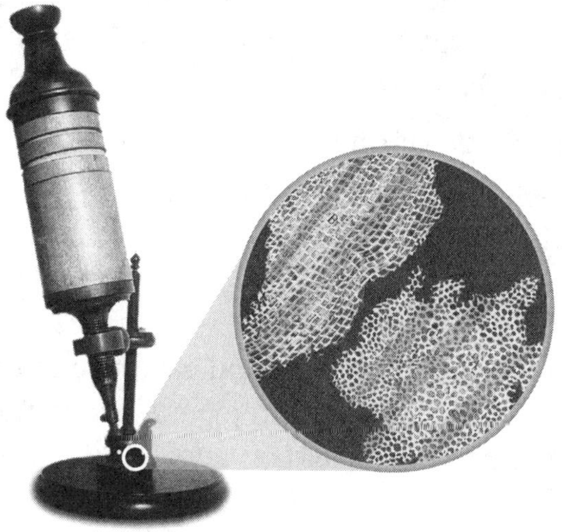

The first cells that Hooke saw were from cork. These cells were easy to see because plant cells have cell walls. At first, Hooke didn't think animals had cells because he couldn't see them. Today we know that all living things are made of cells.

STANDARDS CHECK

LS 1b All organisms are composed of cells—the fundamental unit of life. Most organisms are single cells; other organisms, including humans, are multicellular.

1. Identify What is the basic unit of all living things?

____

In the late 1600s, a Dutch merchant named Anton van Leeuwenhoek studied many different kinds of cells. He made his own microscopes. With them, he looked at tiny pond organisms called protists. He also looked at blood cells, yeasts, and bacteria.

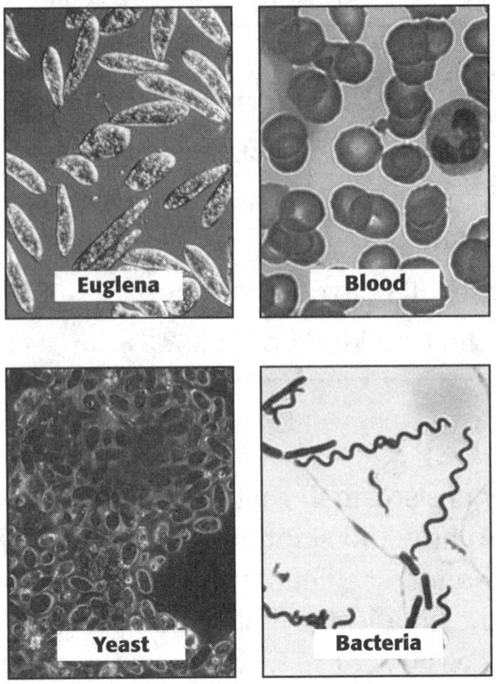
Euglena
Blood
Yeast
Bacteria

Leeuwenhoek looked at many different kinds of cells with his microscope. He was the first person to see bacteria. Bacterial cells are usually much smaller than most other types of cells.

TAKE A LOOK

2. Identify Which of these cells is probably the smallest? Explain your answer.

____

____

____

What Is the Cell Theory?

Since Hooke first saw cork cells, many discoveries have been made about cells. Cells from different organisms can be very different from one another. Even cells from different parts of the same organism can be very different. However, all cells have several important things in common. These observations are known as the *cell theory.* The cell theory has three parts:

1. All organisms are made of one or more cells.
2. The cell is the basic unit of all living things.
3. All cells come from existing cells.

What Are the Parts of a Cell?

Cells come in many shapes and sizes and can have different functions. However, all cells have three parts in common: a cell membrane, genetic material, and organelles. ☑

READING CHECK

3. List What three parts do all cells have in common?

____

____

____

CELL MEMBRANE

All cells are surrounded by a cell membrane. The **cell membrane** is a layer that covers and protects the cell. The membrane separates the cell from its surroundings. The cell membrane also controls all material going in and out of the cell. Inside the cell is a fluid called *cytoplasm.*

GENETIC MATERIAL

All cells contain DNA (deoxyribonucleic acid) at some point in their lives. *DNA* is the genetic material that carries information needed to make proteins, new cells, and new organisms. DNA is passed from parent cells to new cells and it controls the activities of the cell.

The DNA in some cells is found inside a structure called the **nucleus**. Most of your cells have a nucleus.

STANDARDS CHECK

LS 2c Every organism requires a set of instructions for specifying its traits. Heredity is the passage of these instructions from one generation to another.

Word Help: specify
to describe or define in detail

4. Explain What is the function of DNA?

____

____

____

____

____

ORGANELLES

Cells have structures called **organelles** that do different jobs for the cell. Most organelles have a membrane covering them. Different types of cells can have different organelles.

Parts of a Cell

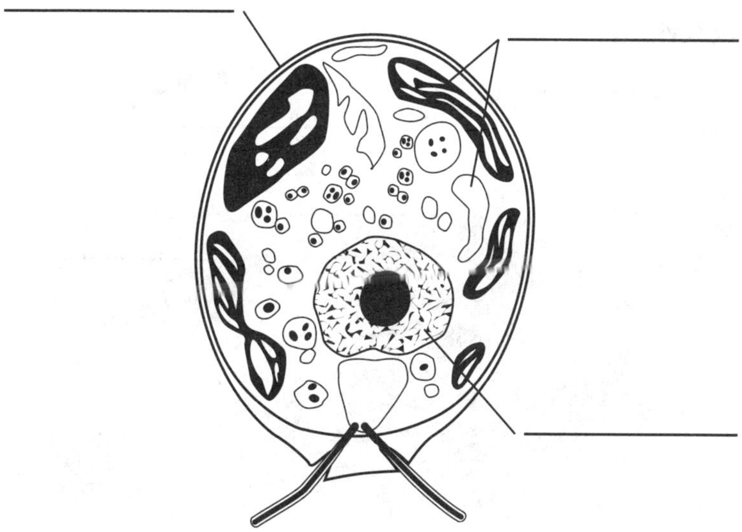

TAKE A LOOK

5. Identify Use the following words to fill in the blank labels on the figure: DNA, cell membrane, organelles.

What Are the Two Kinds of Cells?

There are two basic kinds of cells—cells with a nucleus and cells without a nucleus. Those without a nucleus are called *prokaryotic cells*. Those with a nucleus are called *eukaryotic cells*. ☑

READING CHECK

6. Compare What is one way prokaryotic and eukaryotic cells differ?

____

____

____

What Are Prokaryotes?

A **prokaryote** is an organism made of one cell that does not have a nucleus or other organelles covered by a membrane. Prokaryotes are made of prokaryotic cells. There are two types of prokaryotes: bacteria and archaea.

Critical Thinking

7. Make Inferences Why do you think bacteria can live in your digestive system without making you sick?

____

____

____

____

BACTERIA

The most common prokaryotes are bacteria (singular, *bacterium*). Bacteria are the smallest known cells. These tiny organisms live almost everywhere. Some bacteria live in the soil and water. Others live on or inside other organisms. You have bacteria living on your skin and teeth and in your digestive system. The following are some characteristics of bacteria:

- no nucleus
- circular DNA shaped like a twisted rubber band
- no membrane-covered (or *membrane-bound*) organelles
- a cell wall outside the cell membrane
- a *flagellum* (plural, *flagella*), a tail-like structure that some bacteria use to help them move

A Bacterium

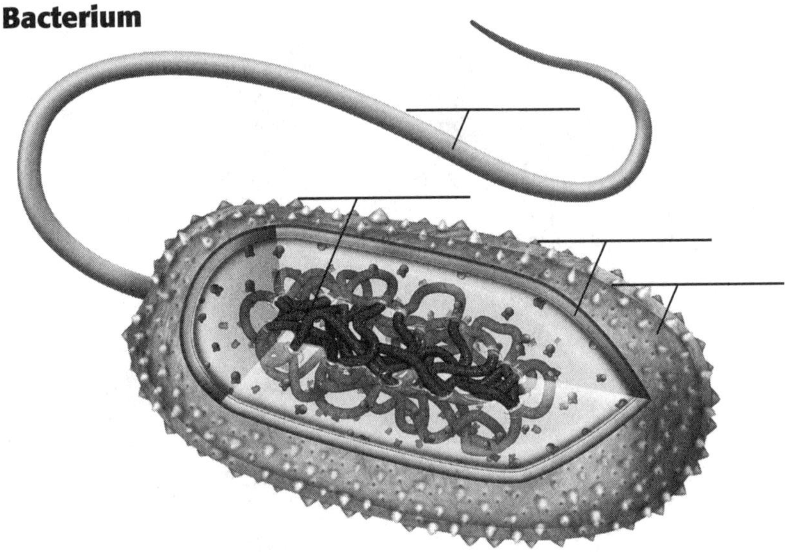

TAKE A LOOK

8. Identify Label the parts of the bacterium using the following terms: DNA, flagellum, cell membrane, cell wall.

ARCHAEA

Archaea (singular, *archaeon*) and bacteria share the following characteristics:

- no nucleus
- no membrane-bound organelles
- circular DNA
- a cell wall

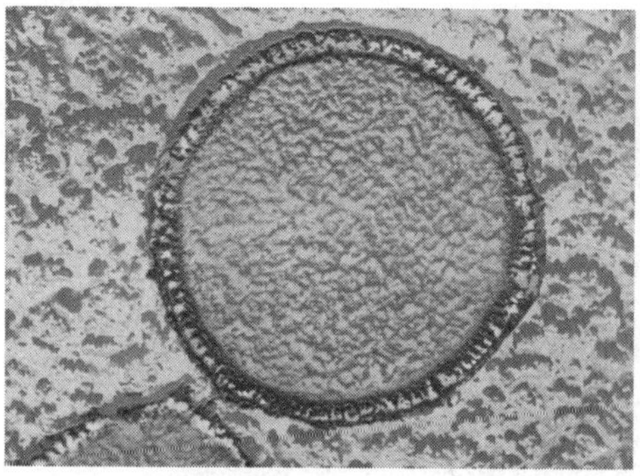

This photograph was taken with an electron microscope. This archaeon lives in volcanic vents deep in the ocean. Temperatures at these vents are very high. Most other living things could not survive there.

Archaea have some other features that no other cells have. For example, the cell wall and cell membrane of archaea are made of different substances from those of bacteria. Some archaea live in places where no other organisms could live. For example, some can live in the boiling water of hot springs. Others can live in toxic places such as volcanic vents filled with sulfur. Still others can live in very salty water in places such as the Dead Sea. ☑

READING CHECK

9. Compare Name two ways that archaea differ from bacteria.

____

____

____

____

____

What Are Eukaryotes?

Eukaryotic cells are the largest cells. They are about 10 times larger than bacteria cells. However, you still need a microscope to see most eukaryotic cells.

Eukaryotes are organisms made of eukaryotic cells. These organisms can have one cell or many cells. Yeast, which makes bread rise, is an example of a eukaryote with one cell. Multicellular organisms, or those made of many cells, include plants and animals.

Unlike prokaryotic cells, eukaryotic cells have a nucleus that holds their DNA. Eukaryotic cells also have membrane-bound organelles. ☑

READING CHECK

10. Identify Name two things eukaryotic cells have that prokaryotic cells do not.

____

____

Eukaryotic Cell

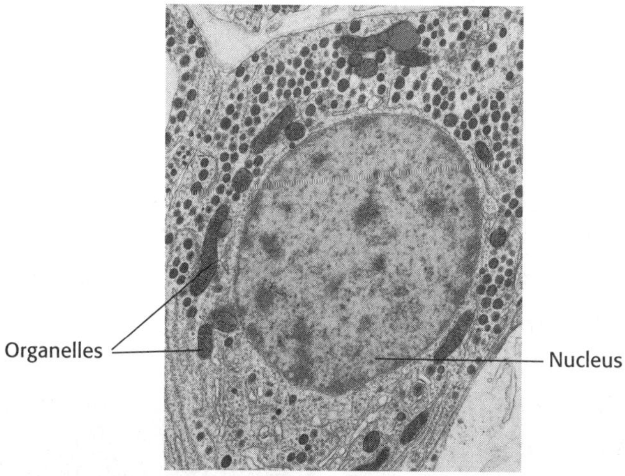
Organelles
Nucleus

Organelles in a Typical Eukaryotic Cell

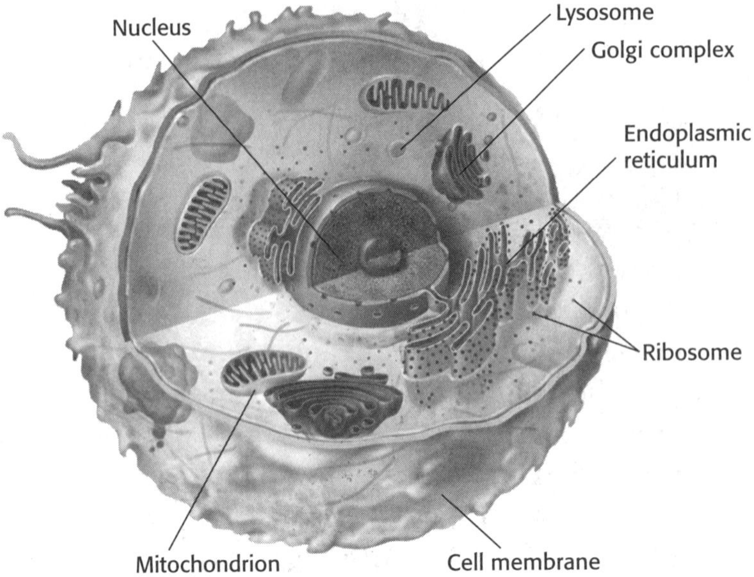
Nucleus
Lysosome
Golgi complex
Endoplasmic reticulum
Ribosome
Mitochondrion
Cell membrane

TAKE A LOOK

11. Identify Where is the genetic material found in this cell?

____

Why Are Cells So Small?

Your body is made of trillions of cells. Most cells are so small you need a microscope to see them. More than 50 human cells can fit on the dot of this letter *i*. However, some cells are big. For example, the yolk of a chicken egg is one big cell! Why, then, are most cells small?

Critical Thinking

12. Apply Concepts The yolk of a chicken egg is a very large cell. Unlike most cells, egg yolks do not have to take in any nutrients. Why does this allow the cell to be so big?

____

____

____

____

____

____

____

Cells take in food and get rid of waste through their outer surfaces. As a cell gets larger, it needs more food to survive. It also produces more waste. This means that more materials have to pass through the surface of a large cell than a small cell.

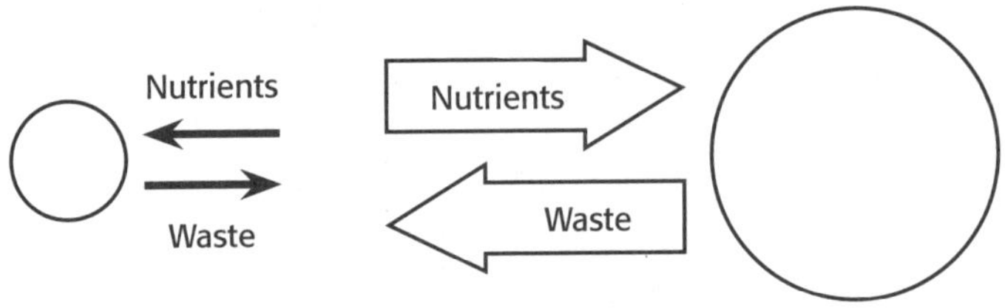
Nutrients
Waste
Nutrients
Waste

Large cells have to take in more nutrients and get rid of more wastes than small cells.

As a cell's volume increases, its outside surface area grows too. However, volume always grows faster than surface area. If the cell volume gets too big, the surface area will not be large enough for the cell to survive. The cell will not be able to take in enough nutrients or get rid of all its wastes. This means that surface area limits the size of most cells.

SURFACE AREA AND VOLUME OF CELLS

To understand how surface area limits the size of a cell, study the figures below. Imagine that the cubes are cells. You can calculate the surface areas and volumes of the cells using these equations:

$$\textit{volume of cube} = \textit{side} \times \textit{side} \times \textit{side}$$

$$\textit{surface area of cube} = \textit{number of sides} \times \textit{area of side}$$

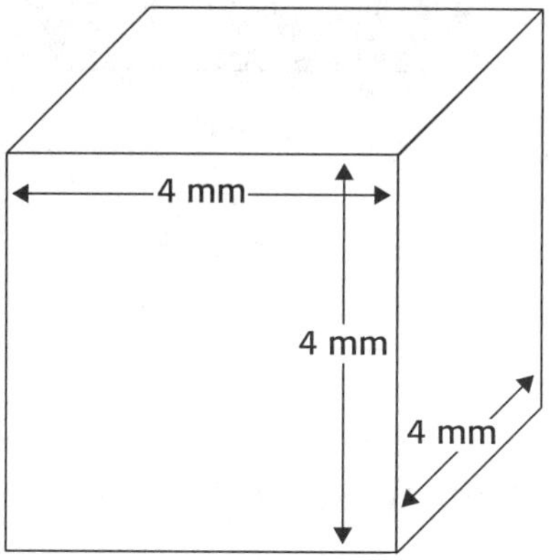
4 mm
4 mm
4 mm

The volume of this cell is 64 mm^3. Its surface area is 96 mm^2.

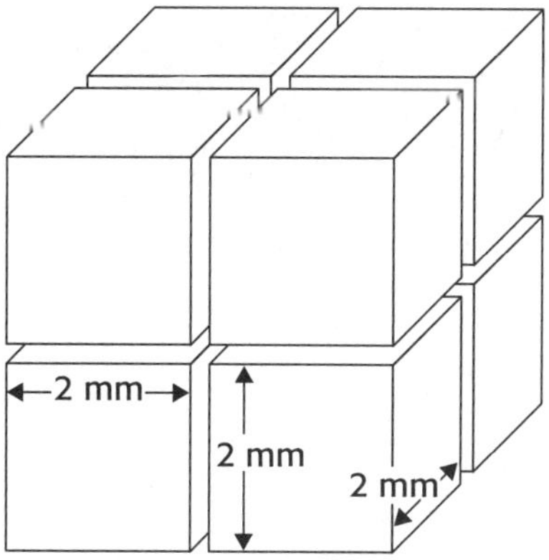
2 mm
2 mm
2 mm

When the large cell is broken up into 8 smaller cells, the total volume stays the same. However, all of the small cells together have more surface area than the large cell. The total surface area of the small cells is 192 mm^2.

Math Focus

13. Calculate Ratios Scientists say that most cells are small because of the surface area-to-volume ratio. What is this ratio for the large cell?

____

TAKE A LOOK

14. Compare Which cell has a greater surface area compared to its volume—the large cell or one of the smaller cells?

____

The large cell takes in and gets rid of the same amount of material as all of the smaller cells. However, the large cell does not have as much surface area as the smaller cells. Therefore, it cannot take in nutrients or get rid of wastes as easily as each of the smaller cells.

Name ______________________ Class ______________ Date ______________

Section 1 Review

NSES **LS 1a, 1b, 1c, 2c, 3b, 5a**

SECTION VOCABULARY

cell in biology, the smallest unit that can perform all life processes; cells are covered by a membrane and have DNA and cytoplasm

cell membrane a phospholipid layer that covers a cell's surface; acts as a barrier between the inside of a cell and the cell's environment

eukaryote an organism made up of cells that have a nucleus enclosed by a membrane; eukaryotes include animals, plants, and fungi, but not archaea or bacteria

nucleus in a eukaryotic cell, a membrane-bound organelle that contains the cell's DNA and that has a role in processes such as growth, metabolism, and reproduction

organelle one of the small bodies in a cell's cytoplasm that are specialized to perform a specific function

prokaryote an organism that consists of a single cell that does not have a nucleus

1. Identify What are the three parts of the cell theory?

____

____

____

2. Compare Fill in the Venn Diagram below to compare prokaryotes and eukaryotes. Be sure to label the circles.

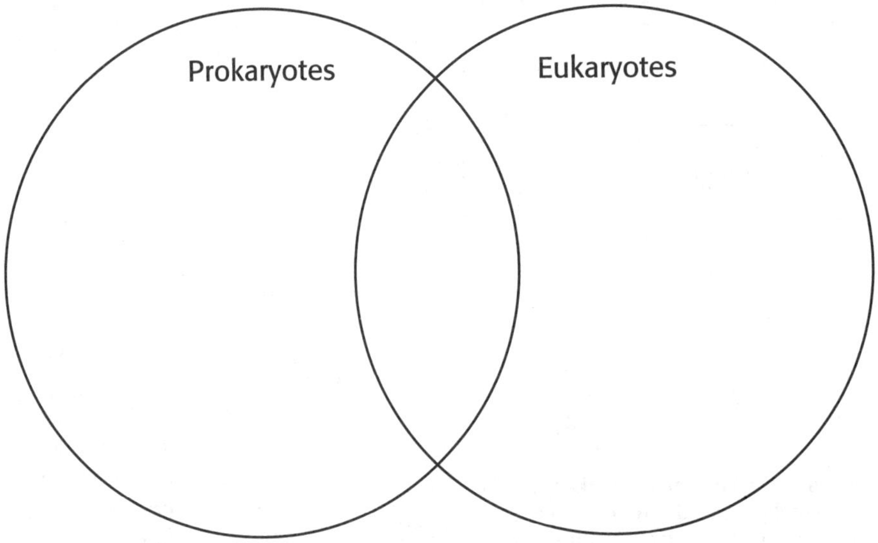
Prokaryotes
Eukaryotes

3. Apply Concepts You have just discovered a new organism. It has only one cell and was found on the ocean floor, at a vent of boiling hot water. The organism has a cell wall but no nucleus. Explain how you would classify this organism.

____

____

____

Name ______________________ Class ______________ Date ______________

SECTION 2

Eukaryotic Cells

BEFORE YOU READ

After you read this section, you should be able to answer these questions:

- What are the parts of a eukaryotic cell?
- What is the function of each part of a eukaryotic cell?

National Science Education Standards
LS 1a, 1b, 1c, 3a, 5a

What Are the Parts of a Eukaryotic Cell?

Plant cells and animal cells are two types of eukaryotic cells. A eukaryotic cell has many parts that help the cell stay alive.

STUDY TIP
Organize As you read this section, make a chart comparing plant cells and animal cells.

CELL WALL

All plant cells have a cell wall. The **cell wall** is a stiff structure that supports the cell and surrounds the cell membrane. The cell wall of a plant cell is made of a type of sugar called cellulose.

Fungi (singular *fungus*), such as yeasts and mushrooms, also have cell walls. The cell walls of fungi are made of a sugar called *chitin*. Prokaryotic cells such as bacteria and archaea also have cell walls. ☑

READING CHECK
1. Identify Name two kinds of eukaryotes that have a cell wall.

____

____

Plant Cell

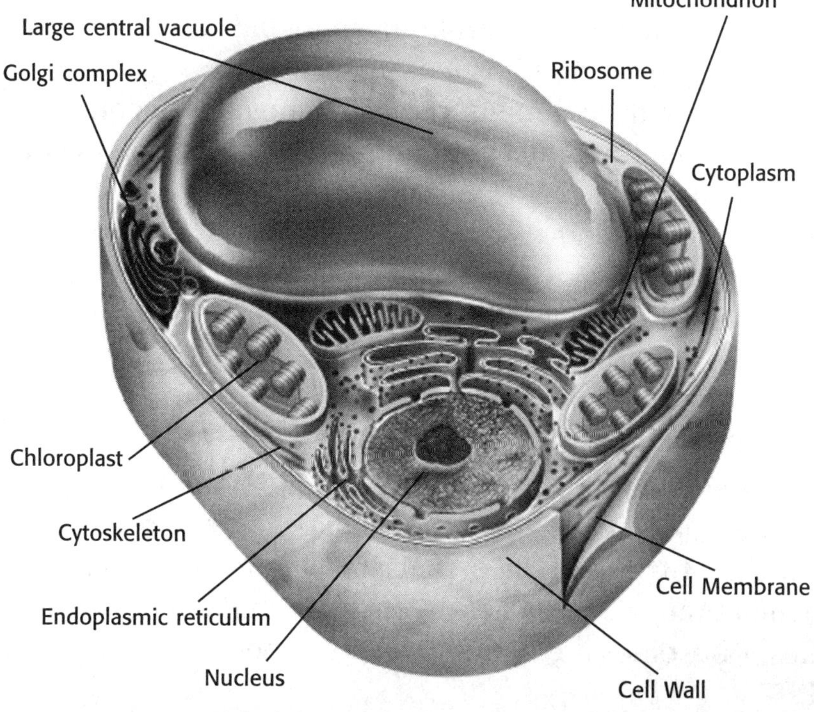
Large central vacuole
Golgi complex
Mitochondrion
Ribosome
Cytoplasm
Chloroplast
Cytoskeleton
Endoplasmic reticulum
Nucleus
Cell Membrane
Cell Wall

TAKE A LOOK
2. Identify Describe where the cell wall is located.

____

____

Name ______________________ Class ______________ Date ______________

Animal Cell

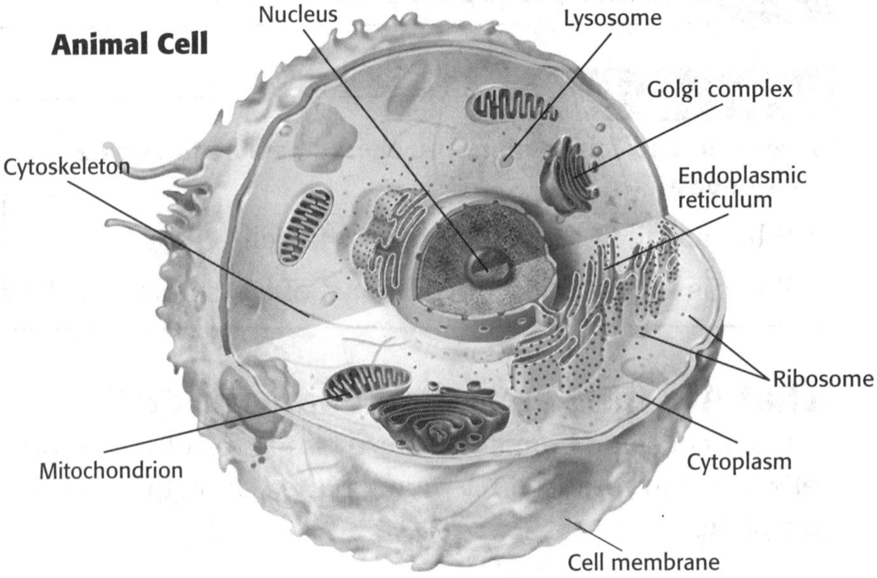
Nucleus
Lysosome
Golgi complex
Cytoskeleton
Endoplasmic reticulum
Ribosome
Mitochondrion
Cytoplasm
Cell membrane

TAKE A LOOK

3. Compare Compare the pictures of an animal cell and a plant cell. Name three structures found in both.

____

____

____

STANDARDS CHECK

LS 1c Cells carry on the many functions needed to sustain life. They grow and divide, thereby producing more cells. This requires that they take in nutrients, which they use to provide energy for the work that cells do and to make the materials that a cell or an organism needs.

Word Help: function
use or purpose

4. Explain What is the main function of the cell membrane?

____

____

____

____

____

____

CELL MEMBRANE

All cells have a cell membrane. The cell membrane is a protective barrier that surrounds the cell. It separates the cell from the outside environment. In cells that have a cell wall, the cell membrane is found just inside the cell wall.

The cell membrane is made of different materials. It contains proteins, lipids, and phospholipids. Proteins are molecules made by the cell for a variety of functions. Lipids are compounds that do not dissolve in water. They include fats and cholesterol. Phospholipids are lipids that contain the element phosphorous.

The proteins and lipids in the cell membrane control the movement of materials into and out of the cell. A cell needs materials such as nutrients and water to survive and grow. Nutrients and wastes go in and out of the cell through the proteins in the cell membrane. Water can pass through the cell membrane without the help of proteins.

RIBOSOMES

Ribosomes are organelles that make proteins. They are the smallest organelles. A cell has many ribosomes. Some float freely in the cytoplasm. Others are attached to membranes or to other organelles. Unlike most organelles, ribosomes are not covered by a membrane. ☑

READING CHECK

5. Compare How are ribosomes different from other organelles?

____

____

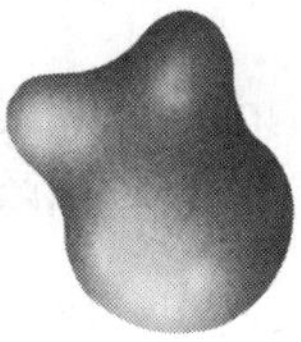

Ribosome
This organelle is where amino acids are hooked together to make proteins.

NUCLEUS

The nucleus is a large organelle in a eukaryotic cell. It contains the cell's genetic material, or DNA. DNA has the instructions that tell a cell how to make proteins.

The nucleus is covered by two membranes. Materials pass through pores in the double membrane. The nucleus of many cells has a dark area called the *nucleolus*.

The Nucleus

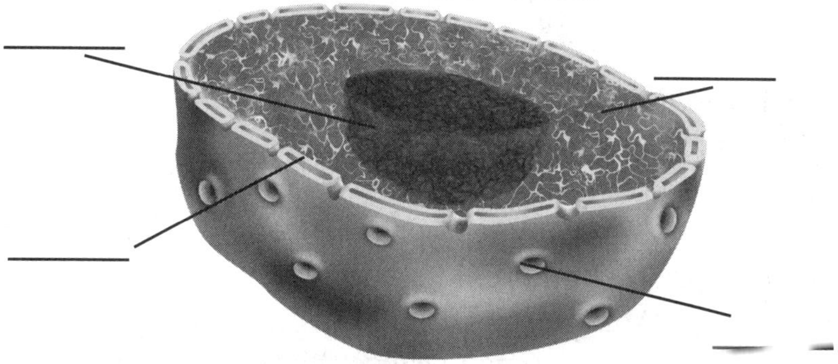

TAKE A LOOK

6. Identify Label the diagram of a nucleus using these terms: pore, DNA, nucleolus, double membrane.

ENDOPLASMIC RETICULUM

Many chemical reactions take place in the cell. Many of these reactions happen on or inside the endoplasmic reticulum. The **endoplasmic reticulum** (ER) is a system of membranes with many folds in which proteins, lipids, and other materials are made.

The ER is also part of the cell's delivery system. Its folds have many tubes and passageways. Materials move through the ER to other parts of the cell.

There are two types of ER: rough and smooth. Smooth ER makes lipids and helps break down materials that could damage the cell. Rough ER has ribosomes attached to it. The ribosomes make proteins. The proteins are then delivered to other parts of the cell by the ER. ☑

READING CHECK

7. Compare What is the difference between smooth ER and rough ER?

____

____

____

____

____

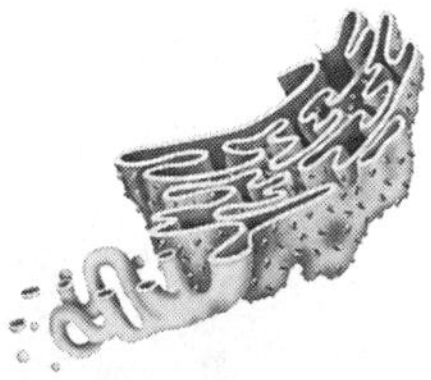

Endoplasmic reticulum
This organelle makes lipids, breaks down drugs and other substances, and packages proteins for the Golgi complex.

MITOCHONDRIA

A **mitochondrion** (plural, *mitochondria*) is the organelle in which sugar is broken down to make energy. It is the main power source for a cell.

A mitochondrion is covered by two membranes. Most of a cell's energy is made in the inside membrane. Energy released by mitochondria is stored in a molecule called ATP. The cell uses ATP to do work.

Mitochondria are about the same size as some bacteria. Like bacteria, mitochondria have their own DNA. The DNA in mitochondria is different from the cell's DNA. ☑

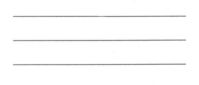

READING CHECK

8. Compare How are mitochondria like bacteria?

____

____

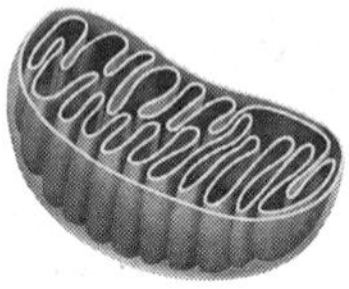

Mitochondrion
This organelle breaks down food molecules to make ATP.

CHLOROPLASTS

Plants and algae have chloroplasts in some of their cells. *Chloroplasts* are organelles in which photosynthesis takes place. *Photosynthesis* is a process by which plants use sunlight, carbon dioxide, and water to make sugar and oxygen. Animal cells do not have chloroplasts.

Chloroplasts are green because they contain a green molecule called *chlorophyll*. Chlorophyll traps the energy of sunlight. Mitochondria then use the sugar made in photosynthesis to make ATP.

Critical Thinking

9. Infer Why don't animal cells need chloroplasts?

____

____

____

____

Chloroplast
This organelle uses the energy of sunlight to make food.

CYTOSKELETON

The cytoskeleton is a web of proteins inside the cell. It acts as both a skeleton and a muscle. The cytoskeleton helps the cell keep its shape. It also helps some cells, such as bacteria, to move.

VESICLES

A **vesicle** is a small sac that surrounds material to be moved. The vesicle moves material to other areas of the cell or into or out of the cell. All eukaryotic cells have vesicles.

GOLGI COMPLEX

The **Golgi complex** is the organelle that packages and distributes proteins. It is the "post office" of the cell. The Golgi complex looks like the smooth ER.

The ER delivers lipids and proteins to the Golgi complex. The Golgi complex can change the lipids and proteins to do different jobs. The final products are then enclosed in a piece of the Golgi complex's membrane. This membrane pinches off to form a vesicle. The vesicle transports the materials to other parts of the cell or out of the cell. ☑

Golgi complex
This organelle processes and transports proteins and other materials out of cell.

READING CHECK

10. Define What is the function of the Golgi complex?

____

____

____

LYSOSOMES

Lysosomes are organelles that contain digestive enzymes. The enzymes destroy worn-out or damaged organelles, wastes, and invading particles.

Lysosomes are found mainly in animal cells. The cell wraps itself around a particle and encloses it in a vesicle. Lysosomes bump into the vesicle and pour enzymes into it. The enzymes break down the particles inside the vesicle. Without lysosomes, old or dangerous materials could build up and damage or kill the cell.

Lysosome
This organelle digests food particles, wastes, cell parts, and foreign invaders.

VACUOLES

A vacuole is a vesicle. In plant and fungal cells, some vacuoles act like lysosomes. They contain enzymes that help a cell digest particles. The large central vacuole in plant cells stores water and other liquids. Large vacuoles full of water help support the cell. Some plants wilt when their vacuoles lose water. ☑

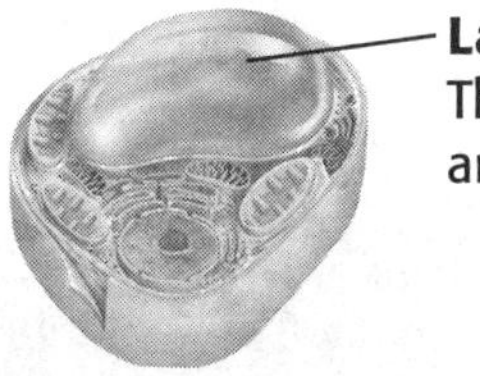

Large central vacuole
This organelle stores water and other materials.

READING CHECK

11. Identify Vacuoles are found in what types of eukaryotic cells?

____

____

Name ________________________ Class ______________ Date ______________

Section 2 Review

NSES **LS 1a, 1b, 1c, 3a, 5a**

SECTION VOCABULARY

cell wall a rigid structure that surrounds the cell membrane and provides support to the cell

endoplasmic reticulum a system of membranes that is found in a cell's cytoplasm and that assists in the production, processing, and transport of proteins and in the production of lipids

Golgi complex cell organelle that helps make and package materials to be transported out of the cell

lysosome a cell organelle that contains digestive enzymes

mitochondrion in eukaryotic cells, the cell organelle that is surrounded by two membranes and that is the site of cellular respiration

ribosome cell organelle composed of RNA and protein; the site of protein synthesis

vesicle a small cavity or sac that contains materials in a eukaryotic cell

1. Compare Name three parts of a plant cell that are not found in an animal cell.

____

____

2. Explain How does a cell get water and nutrients?

____

____

____

3. Explain What would happen to an animal cell if it had no lysosomes?

____

____

____

4. Apply Concepts Which kind of cell in the human body do you think would have more mitochondria—a muscle cell or a skin cell? Explain.

____

____

____

____

5. List What are two functions of the cytoskeleton?

____

____

Name ______________________ Class ______________ Date ______________

CHAPTER 1 Cells: The Basic Units of Life

SECTION 3

The Organization of Living Things

BEFORE YOU READ

After you read this section, you should be able to answer these questions:

- What are the advantages of being multicellular?
- What are the four levels of organization in living things?
- How are structure and function related in an organism?

National Science Education Standards
LS 1a, 1d, 1e

What Is an Organism?

Anything that can perform life processes by itself is an **organism**. An organism made of a single cell is called a *unicellular organism*. An organism made of many cells is a multicellular organism. The cells in a multicellular organism depend on each other for the organism to survive. ☑

STUDY TIP

Outline As you read, make an outline of this section. Use the heading questions from the section in your outline.

What Are the Benefits of Having Many Cells?

Some organisms exist as one cell. Others can be made of trillions of cells. A *multicellular organism* is an organism made of many cells.

There are three benefits of being multicellular: larger size, longer life, and specialization of cells.

READING CHECK

1. Define What is an organism?

____

____

____

LARGER SIZE

Most multicellular organisms are bigger than one-celled organisms. In general, a larger organism, such as an elephant, has few predators. ☑

READING CHECK

2. Identify Name one way that being large can benefit an organism.

____

____

____

LONGER LIFE

A multicellular organism usually lives longer than a one-celled organism. A one-celled organism is limited to the life span of its one cell. The life span of a multicellular organism, however, is not limited to the life span of any one of its cells.

SPECIALIZATION

In a multicellular organism, each type of cell has a particular job. Each cell does not have to do everything the organism needs. Specialization makes the organism more efficient.

Standards Check

LS 1d Specialized cells perform specialized functions in multicellular organisms. Groups of specialized cells cooperate to form a tissue, such as a muscle. Different tissues are in turn grouped together and form larger functional units, called organs. Each type of cell, tissue, and organ has a distinct structure and set of functions that serve the organism as a whole.

3. List What are the four levels or organization for an organism?

____

____

____

____

What Are the Four Levels of Organization of Living Things?

Multicellular organisms have four levels of organization:

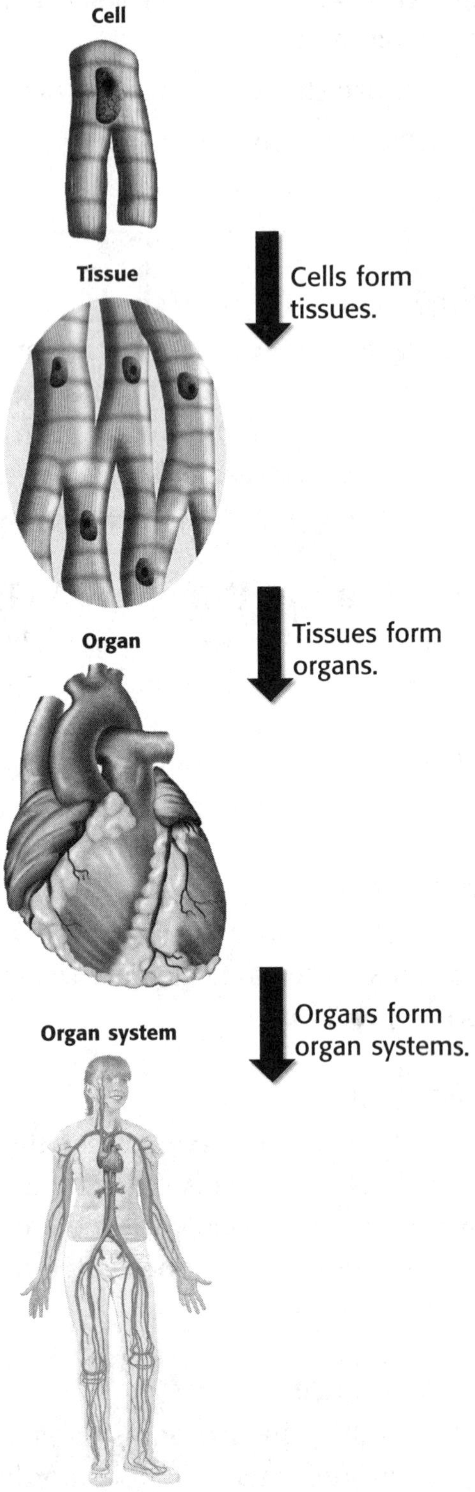
Cell
Tissue
Cells form tissues.
Organ
Tissues form organs.
Organ system
Organs form organ systems.

Organ systems form organisms such as you.

TAKE A LOOK

4. Explain Are the cells that make up heart tissue prokaryotic or eukaryotic? How do you know?

____

____

____

____

CELLS WORK TOGETHER AS TISSUES

A **tissue** is a group of cells that work together to perform a specific job. Heart muscle tissue, for example, is made of many heart muscle cells.

Critical Thinking

5. Apply Concepts Do prokaryotes have tissues? Explain.

____

____

____

____

____

TISSUES WORK TOGETHER AS ORGANS

A structure made of two or more tissues that work together to do a certain job is called an **organ**. Your heart, for example, is an organ made of different tissues. The heart has muscle tissues and nerve tissues that work together.

ORGANS WORK TOGETHER AS ORGAN SYSTEMS

A group of organs working together to do a job is called an **organ system**. An example of an organ system is your digestive system. Organ systems depend on each other to help the organism function. For example, the digestive system depends on the cardiovascular and respiratory systems for oxygen.

Say It

Name With a partner, name as many of the organs in the human body as you can.

HOW DOES STRUCTURE RELATE TO FUNCTION?

In an organism, the structure and function of part are related. **Function** is the job the part does. **Structure** is the arrangement of parts in an organism. It includes the shape of a part or the material the part is made of.

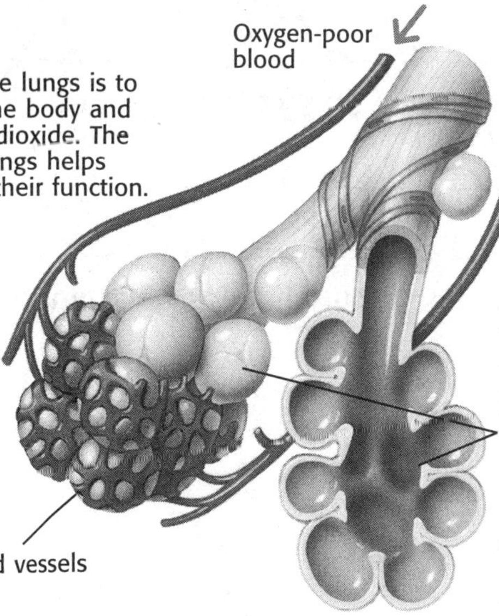
The function of the lungs is to bring oxygen to the body and get rid of carbon dioxide. The structure of the lungs helps them to perform their function.
Oxygen-poor blood
Oxygen-rich blood
The lungs contain tiny, spongy sacs that blood can flow through. Carbon dioxide moves out of the blood and into the sacs. Oxygen flows from the sacs into the blood. If the lungs didn't have this structure, it would be hard for them to perform their function.
Blood vessels

Name ______________________ Class ______________ Date ______________

Section 3 Review

NSES LS 1a, 1d, 1e

SECTION VOCABULARY

function the special, normal, or proper activity of an organ or part

organ a collection of tissues that carry out a specialized function of the body

organ system a group of organisms that work together to perform body functions

organism a living thing; anything that can carry out life processes independently

structure the arrangement of parts in an organism

tissue a group of similar cells that perform a common function

1. List What are three benefits of being multicellular?

____

____

____

2. Apply Concepts Could an organism have organs but no tissues? Explain.

____

____

3. Compare How are structure and function different?

____

____

____

4. Explain What does "specialization of cells" mean?

____

____

5. Apply Concepts Why couldn't your heart have only cardiac tissue?

____

____

6. Explain Why do multicellular organisms generally live longer than unicellular organisms?

____

____

Name ______________________ Class ______________ Date ______________

CHAPTER 2 The Cell in Action

SECTION 1

Exchange with the Environment

BEFORE YOU READ

After you read this section, you should be able to answer these questions:

- How do cells take in food and get rid of wastes?
- What is diffusion?

National Science Education Standards
LS 1a, 1c

Where Do Cells Get the Materials They Need?

What would happen to a factory if its power were shut off or its supply of materials never arrived? What would happen if the factory couldn't get rid of its garbage? Like a factory, an organism must be able to get energy and raw materials and get rid of wastes. These jobs are done by an organism's cells. Materials move in and out of the cell across the cell membrane. Many materials, such as water and oxygen, can cross the membrane by diffusion.

STUDY TIP

Compare As you read, make a chart comparing diffusion and osmosis. In your chart, show how they are similar and how they are different.

What Is Diffusion?

The figure below shows what happens when dye is placed on top of a layer of gelatin. Over time, the dye mixes with the gelatin. Why does this happen?

Everything, including the gelatin and the dye, is made of tiny moving particles. Particles tend to move from places where they are crowded to places where they are less crowded. When there are many of one type of particle, this is a high concentration. When there are fewer of one kind of particle, this is a low concentration. The movement from areas of high concentration to areas of low concentration is called **diffusion**. ☑

READING CHECK

1. Define What is diffusion?

____

____

____

____

At first, the dye and the gelatin are separate from each other.
Dye
Gelatin
After a while, the particles in the dye move into the gelatin. This process is called diffusion.

TAKE A LOOK

2. Identify How do dye particles move through the water?

____

Name ______________________ Class ______________ Date ______________

DIFFUSION OF WATER

Substances, such as water, are made up of particles called *molecules*. Pure water has the highest concentration of water molecules. This means that 100% of the molecules are water molecules. If you mix another substance, such as food coloring, into the water, you lower the concentration of water molecules. This means that water molecules no longer make up 100% of the total molecules.

The figure below shows a container that has been divided by a membrane. The membrane is *semipermeable*—that is, only some substances can pass through it. The membrane lets smaller molecules, such as water, pass through. Larger molecules, such as food coloring, cannot pass through. Water molecules will move across the membrane. The diffusion of water through a membrane is called **osmosis**.

Critical Thinking

3. Apply Concepts Which of the following has a higher concentration of water molecules—200 molecules of water, or a mixture of 300 molecules of water and 100 molecules of food coloring? Explain your answer.

____

____

____

____

____

____

Osmosis

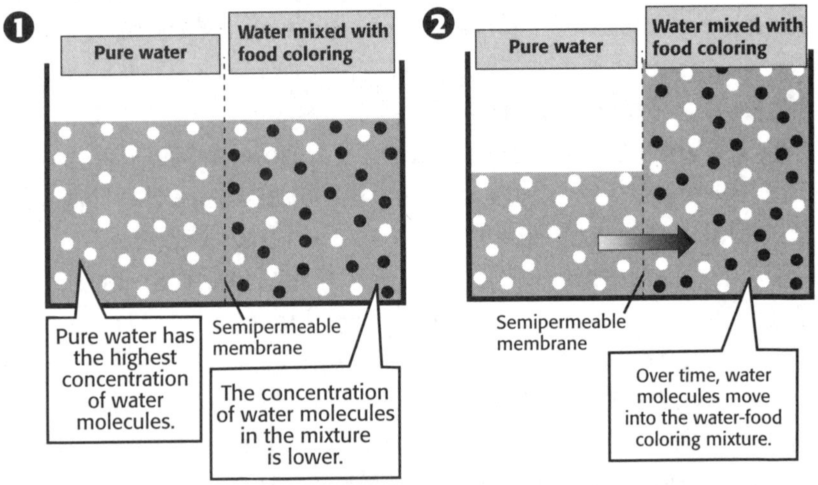

1
Pure water
Water mixed with food coloring
Pure water has the highest concentration of water molecules.
Semipermeable membrane
The concentration of water molecules in the mixture is lower.
2
Pure water
Water mixed with food coloring
Semipermeable membrane
Over time, water molecules move into the water-food coloring mixture.

TAKE A LOOK

4. Explain Why does the volume of liquid in the right-hand side of the container increase with time?

____

____

A cell membrane is a type of semipermeable membrane. This means that water can pass through the cell membrane, but most other substances cannot. The cells of organisms are surrounded by and filled with fluids. These fluids are made mostly of water. Water moves in and out of a cell by osmosis. ☑

READING CHECK

5. Identify How does water move into and out of cells?

____

How Do Small Particles Enter and Leave a Cell?

Small particles, such as sugars, can cross the cell membrane through passageways called *channels*. These channels in the cell membrane are made of proteins. Particles can travel through these channels by passive transport or by active transport.

During **passive transport**, particles move through the cell membrane without using energy from the cell. During passive transport, particles move from areas of high concentration to areas of lower concentration. Diffusion and osmosis are examples of passive transport.

During **active transport**, the cell has to use energy to move particles through channels. During active transport, particles usually move from areas of low concentration to areas of high concentration. ☑

READING CHECK

6. Identify What is needed to move particles from areas of low concentration to areas of high concentration?

____

How Do Large Particles Enter and Leave a Cell?

Large particles cannot move across a cell membrane in the same ways as small particles. Larger particles must move in and out of the cell by endocytosis and exocytosis. Both processes require energy from the cell.

Endocytosis happens when a cell surrounds a large particle and encloses it in a vesicle. A *vesicle* is a sac formed from a piece of cell membrane.

Endocytosis

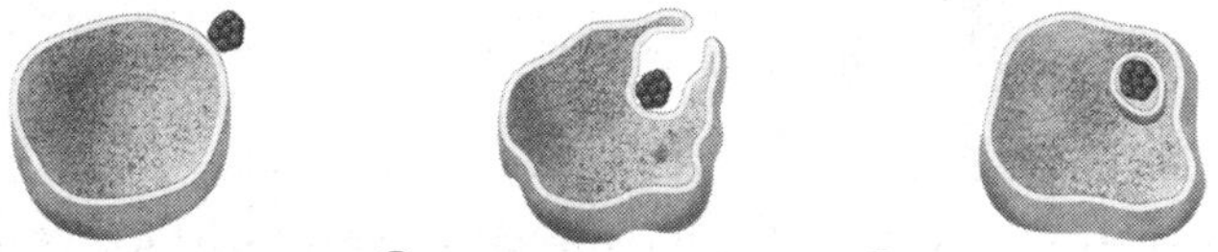

❶ The cell comes into contact with a particle.
❷ The cell membrane begins to wrap around the particle.
❸ Once the particle is completely surrounded, a vesicle pinches off.

TAKE A LOOK

7. Identify Label the vesicle in the figure.

Exocytosis happens when a cell uses a vesicle to move a particle from within the cell to outside the cell. Exocytosis is how cells get rid of large waste particles.

Exocytosis

❶ Large particles that must leave the cell are packaged in vesicles.
❷ The vesicle travels to the cell membrane and fuses with it.
❸ The cell releases the particle to the outside of the cell.

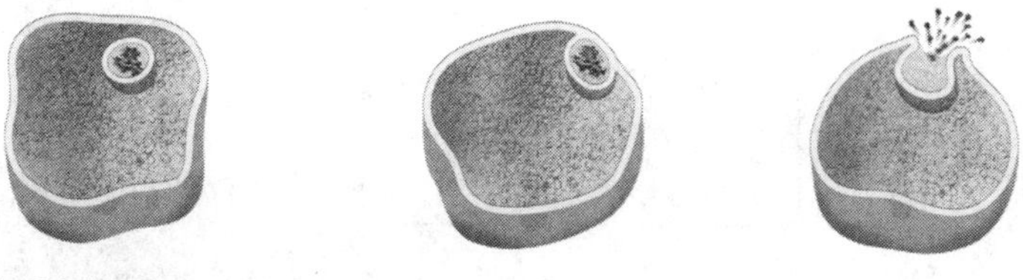

Section 1 Review

NSES LS 1a, 1c

SECTION VOCABULARY

active transport the movement of substances across the cell membrane that requires the cell to use energy

diffusion the movement of particles from regions of higher density to regions of lower density

endocytosis the process by which a cell membrane surrounds a particle and encloses the particle in a vesicle to bring the particle into the cell

exocytosis the process in which a cell releases a particle by enclosing the particle in a vesicle that then moves to the cell surface and fuses with the cell membrane

osmosis the diffusion of water through a semipermeable membrane

passive transport the movement of substances across a cell membrane without the use of energy by the cell

1. **Compare** How is endocytosis different from exocytosis? How are they similar?

____

____

____

2. **Explain** How is osmosis related to diffusion?

____

____

3. **Compare** What are the differences between active and passive transport?

____

____

____

____

4. **Identify** What structures allow small particles to cross cell membranes?

____

5. **Apply Concepts** Draw an arrow in the figure below to show the direction that water molecules will move in.

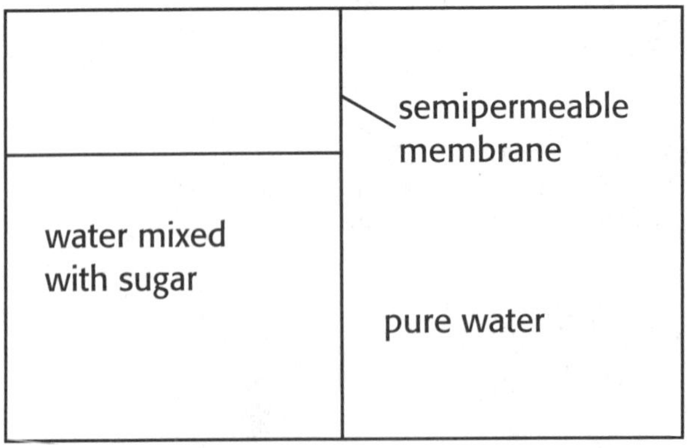
semipermeable membrane
water mixed with sugar
pure water

Name ______________________ Class ____________ Date ____________

CHAPTER 2 The Cell in Action

SECTION 2

Cell Energy

BEFORE YOU READ

After you read this section, you should be able to answer these questions:

- How do plant cells make food?
- How do plant and animal cells get energy from food?

National Science Education Standards
LS 1c, 4c

How Does a Plant Make Food?

The sun is the major source of energy for life on Earth. Plants use carbon dioxide, water, and the sun's energy to make food in a process called **photosynthesis**. The food that plants make gives them energy. When animals eat plants, the plants become sources of energy for the animals.

Plant cells have molecules called *pigments* that absorb light energy. Chlorophyll is the main pigment used in photosynthesis. Chlorophyll is found in chloroplasts. The food plants make is a simple sugar called *glucose*. Photosynthesis also produces oxygen. ☑

STUDY TIP

Compare As you read this section, make a Venn Diagram to compare cellular respiration and fermentation.

READING CHECK

1. Identify In which cell structures does photosynthesis take place?

____

Photosynthesis

$$6CO_2 + 6H_2O + \text{light energy} \rightarrow C_6H_{12}O_6 + 6O_2$$

Carbon dioxide — Water — Glucose — Oxygen

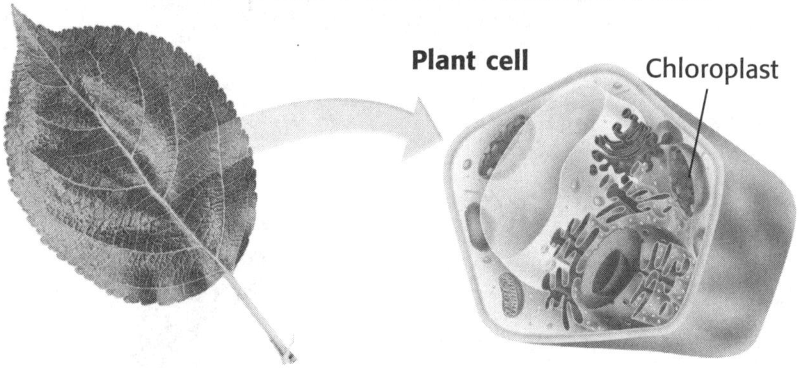
Plant cell
Chloroplast

Photosynthesis takes place in chloroplasts. Chloroplasts are found inside plant cells.

TAKE A LOOK

2. Identify What two materials are produced during photosynthesis?

____

____

How Do Organisms Get Energy from Food?

Both plant and animal cells must break down food molecules to get energy from them. There are two ways cells get energy: cellular respiration and fermentation.

During **cellular respiration**, cells use oxygen to break down food. During **fermentation**, food is broken down without oxygen. Cellular respiration releases more energy from food than fermentation. Most eukaryotes, such as plants and animals, use cellular respiration.

STANDARDS CHECK

LS 1c Cells carry on the many functions needed to sustain life. They grow and divide, thereby producing more cells. This requires that they take in nutrients, which they use to provide energy for the work that cells do and to make the materials a cell or an organism needs.

Word Help: function
to work

Word Help: energy
the capacity to do work

3. Identify Name two ways cells can get energy from food.

____

____

What Happens During Cellular Respiration?

When you hear the word *respiration*, you might think of breathing. However, cellular respiration is different from breathing. Cellular respiration is a chemical process that happens in cells. In eukaryotic cells, such as plant and animal cells, cellular respiration takes place in structures called *mitochondria.*

Recall that to get energy, cells must break down glucose. During cellular respiration, glucose is broken down into carbon dioxide (CO_2) and water (H_2O), and energy is released. This energy is stored in a molecule called *ATP* (adenosine triphosphate). The figure below shows how energy is released when a cow eats grass.

TAKE A LOOK

4. Identify What two materials are needed for cellular respiration?

____

____

5. List What three things are produced during cellular respiration?

____

____

____

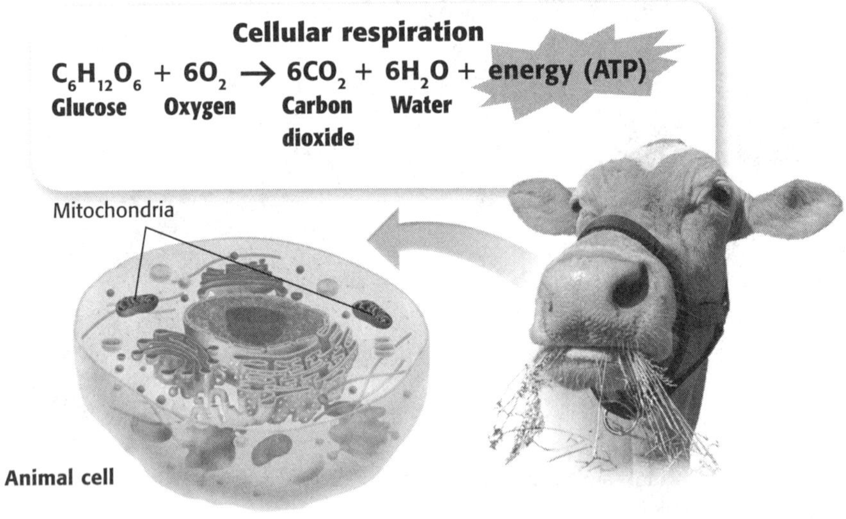
Cellular respiration
$C_6H_{12}O_6 + 6O_2 \rightarrow 6CO_2 + 6H_2O +$ energy (ATP)
Glucose
Oxygen
Carbon dioxide
Water
Mitochondria
Animal cell

The mitochondria in the cells of this cow will use cellular respiration to release the energy stored in the grass.

The Connection Between Photosynthesis and Cellular Respiration

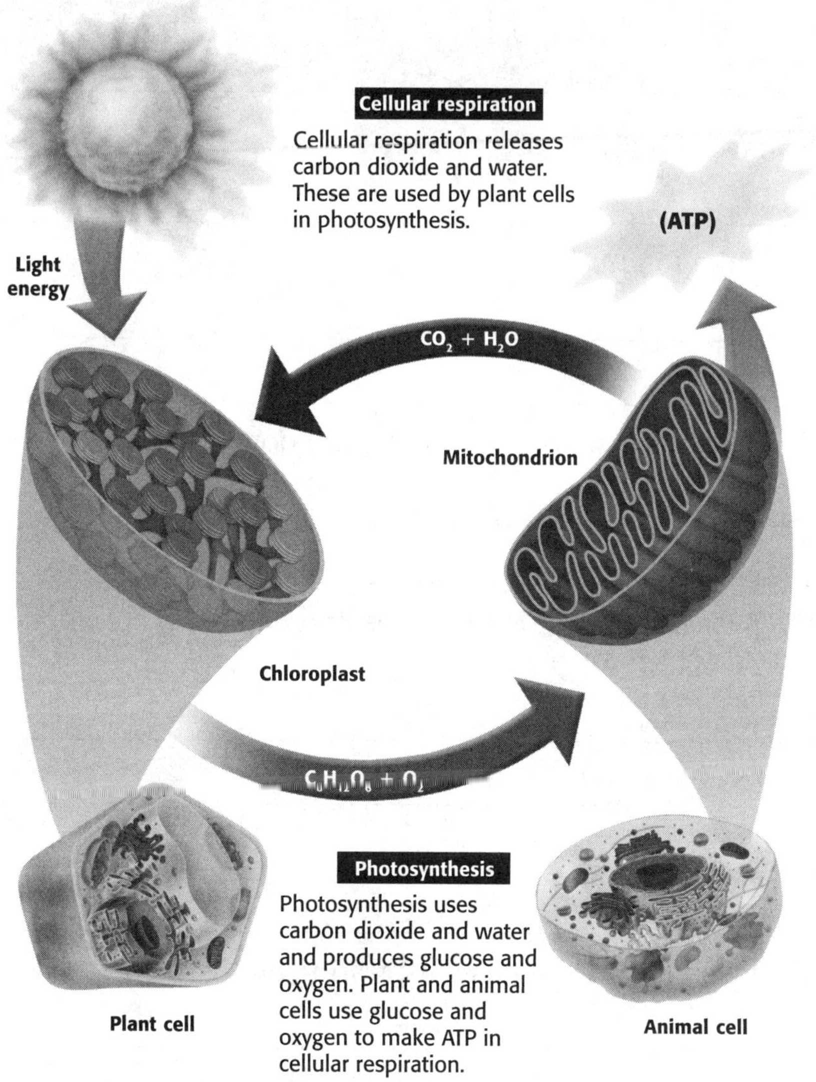
Cellular respiration
Cellular respiration releases carbon dioxide and water. These are used by plant cells in photosynthesis.
(ATP)
Light energy
$CO_2 + H_2O$
Mitochondrion
Chloroplast
$C_6H_{12}O_6 + O_2$
Photosynthesis
Photosynthesis uses carbon dioxide and water and produces glucose and oxygen. Plant and animal cells use glucose and oxygen to make ATP in cellular respiration.
Plant cell
Animal cell

How Is Fermentation Different from Cellular Respiration?

During fermentation, cells break down glucose without oxygen. Some bacteria and fungi rely only on fermentation to release energy from food. However, cells in other organisms may use fermentation when there is not enough oxygen for cellular respiration.

When you exercise, your muscles use up oxygen very quickly. When cells don't have enough oxygen, they must use fermentation to get energy. Fermentation creates a byproduct called *lactic acid*. This is what makes your muscles ache if you exercise too hard or too long.

Critical Thinking

6. Apply Concepts What would happen if oxygen were not produced during photosynthesis?

____

____

____

____

____

____

____

TAKE A LOOK

7. Complete Plant and animal cells use glucose and oxygen to make

______________________________.

Say It

Research Use the school library or the Internet to research an organism that uses fermentation. What kind of organism is it? Where is it found? Is this organism useful to humans? Present your findings to the class.

Name ______________________ Class ______________ Date ______________

Section 2 Review

NSES LS 1c, 4c

SECTION VOCABULARY

cellular respiration the process by which cells use oxygen to produce energy from food **fermentation** the breakdown of food without the use of oxygen	**photosynthesis** the process by which plants, algae, and some bacteria use sunlight, carbon dioxide, and water to make food.

1. Identify What kind of cells have chloroplasts?

____

2. Explain How do plant cells make food?

____

____

3. Explain Why do plant cells need both chloroplasts and mitochondria?

____

____

4. Apply Concepts How do the processes of photosynthesis and cellular respiration work together?

____

____

____

____

5. Compare What is one difference between cellular respiration and fermentation?

____

____

____

____

6. Explain Do your body cells always use cellular respiration to break down glucose? Explain your answer.

____

____

____

Name ______________________ Class ______________ Date ______________

SECTION 3

The Cell Cycle

BEFORE YOU READ

After you read this section, you should be able to answer these questions:

- How are new cells made?
- What is mitosis?
- How is cell division different in animals and plants?

National Science Education Standards
LS 1c, 2d

How Are New Cells Made?

As you grow, you pass through different stages in your life. Cells also pass through different stages in their life cycles. These stages are called the **cell cycle**. The cell cycle starts when a cell is made, and ends when the cell divides to make new cells.

Before a cell divides, it makes a copy of its DNA (deoxyribonucleic acid). *DNA* is a molecule that contains all the instructions for making new cells. The DNA is stored in structures called **chromosomes**. Cells make copies of their chromosomes so that new cells have the same chromosomes as the parent cells. Although all cells pass through a cell cycle, the process differs in prokaryotic and eukaryotic cells. ☑

STUDY TIP
Summarize As you read this section, make a diagram showing the stages of the eukaryotic cell cycle.

READING CHECK
1. Explain What must happen before a cell can divide?

____

____

____

How Do Prokaryotic Cells Divide?

Prokaryotes have only one cell. Prokaryotic cells have no nucleus. They also have no organelles that are surrounded by membranes. The DNA for prokaryotic cells, such as bacteria, is found on one circular chromosome. The cell divides by a process called *binary fission.* During binary fission, the cell splits into two parts. Each part has one copy of the cell's DNA.

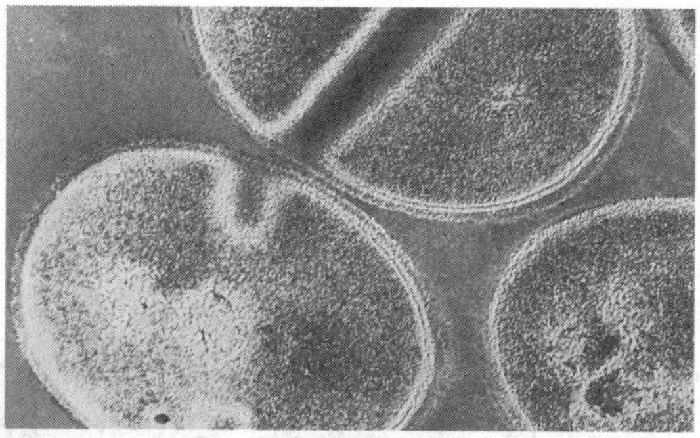

Bacteria reproduce by binary fission.

TAKE A LOOK
2. Complete Prokaryotic cells divide by ______________

______________.

How Do Eukaryotic Cells Divide?

Different kinds of eukaryotes have different numbers of chromosomes. However, complex eukaryotes do not always have more chromosomes than simpler eukaryotes. For example, potatoes have 48 chromosomes, but humans have 46. Many eukaryotes, including humans, have pairs of similar chromosomes. These pairs are called **homologous chromosomes**. One chromosome in a pair comes from each parent.

Cell division in eukaryotic cells is more complex than in prokaryotic cells. The cell cycle of a eukaryotic cell has three stages: interphase, mitosis, and cytokinesis.

The first stage of the cell cycle is called *interphase*. During interphase, the cell grows and makes copies of its chromosomes and organelles. The two copies of a chromosome are called *chromatids*. The two chromatids are held together at the *centromere*.

Critical Thinking

3. Compare What is the difference between a chromosome and a chromatid?

____

____

____

____

____

This duplicated chromosome consists of two chromatids. The chromatids are joined at the centromere.

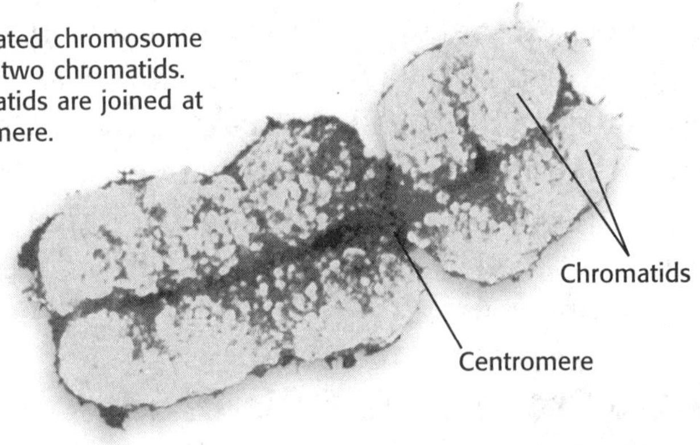
Chromatids
Centromere

The second stage of the cell cycle is called **mitosis**. During this stage, the chromatids separate. This allows each new cell to get a copy of each chromosome. Mitosis happens in four phases, as shown in the figure on the next page: prophase, metaphase, anaphase, and telophase.

The third stage of the cell cycle is called **cytokinesis**. During this stage, the cytoplasm of the cell divides to form two cells. These two cells are called *daughter cells*. The new daughter cells are exactly the same as each other. They are also exactly the same as the original cell. ☑

READING CHECK

4. Identify What are the three stages of the eukaryotic cell cycle?

____

____

____

THE CELL CYCLE

The figure on the following page shows the cell cycle. In this example, the stages of the cell cycle are shown in a eukaryotic cell that has only four chromosomes.

Name ______________________ Class ______________ Date ______________

Interphase Before mitosis begins, chromosomes are copied. Each chromosome is then made of two chromatids.

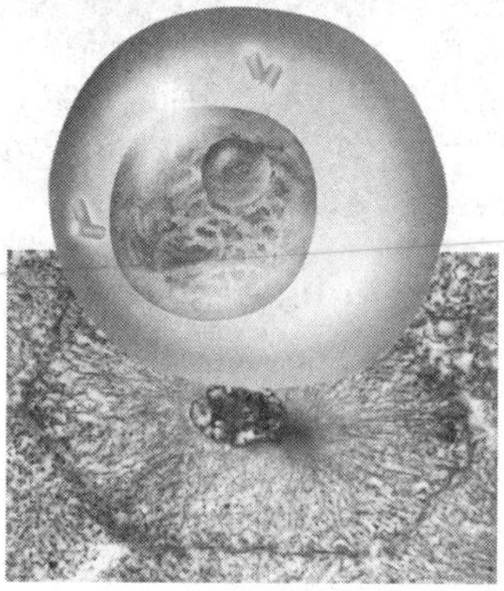

Mitosis Phase 1 (Prophase) Mitosis begins. Chromatids condense from long strands to thick rods.

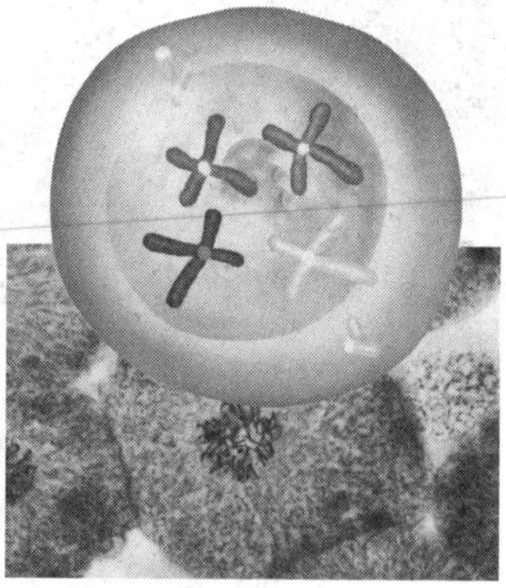

Mitosis Phase 2 (Metaphase) The nuclear membrane dissolves. Chromosome pairs line up around the equator of the cell.

Mitosis Phase 3 (Anaphase) Chromatids separate and move to opposite sides of the cell.

Mitosis Phase 4 (Telophase) A nuclear membrane forms around each set of chromosomes. The chromosomes unwind. Mitosis is complete.

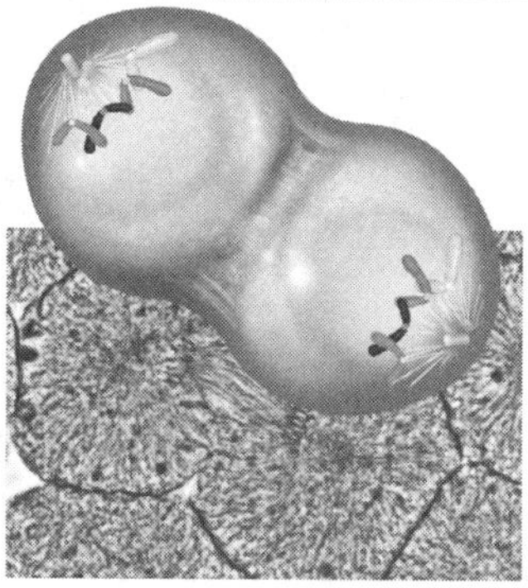

Cytokinesis In cells with no cell wall, the cell pinches in two.

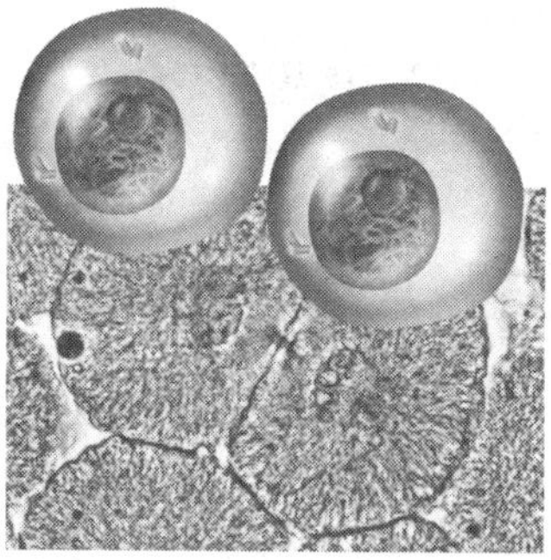

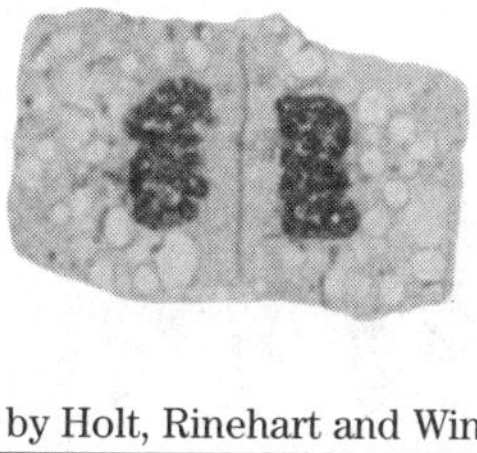

In cells with a cell wall, a cell plate forms and separates the new cells.

Math Focus

5. Calculate Cell A takes 6 h to complete division. Cell B takes 8 h to complete division. After 24 h, how many more copies of cell A than cell B will there be?

____

TAKE A LOOK

6. List What are the four phases of mitosis?

____

____

____

____

7. Identify What structure do plant cells have during cytokinesis that animal cells do not have?

____

Name ______________________ Class ______________ Date ______________

Section 3 Review

NSES LS 1c, 2d

SECTION VOCABULARY

cell cycle the life cycle of a cell

chromosome in a eukaryotic cell, one of the structures in the nucleus that are made up of DNA and protein; in a prokaryotic cell, the main ring of DNA

cytokinesis the division of cytoplasm of a cell

homologous chromosomes chromosomes that have the same sequence of genes and the same structure

mitosis in eukaryotic cells, a process of cell division that forms two new nuclei, each of which has the same number of chromosomes

1. Compare How does the DNA of prokaryotic and eukaryotic cells differ?

____

____

2. Summarize Complete the Process Chart to explain the three stages of the cell cycle. Include the four phases of mitosis.

↓
Mitosis begins with prophase. The chromosomes condense.
↓
↓
During telophase the nuclear membrane forms. The chromosomes lengthen and mitosis ends.
↓

3. Explain Why does a cell make a copy of its DNA before it divides?

____

____

____

4. Infer Why is cell division in eukaryotic cells more complex than in prokaryotic cells?

____

____

____

Name ______________________ Class ______________ Date ______________

CHAPTER 3 Heredity

SECTION 1

Mendel and His Peas

BEFORE YOU READ

After you read this section, you should be able to answer these questions:

- What is heredity?
- How did Gregor Mendel study heredity?

National Science Education Standards
LS 2b, 2e

What Is Heredity?

Why don't you look like a rhinoceros? The answer to that question seems simple. Neither of your parents is a rhinoceros. Only a human can pass on its traits to make another human. Your parents passed some of their traits on to you. The passing of traits from parents to offspring is called **heredity**.

About 150 years ago, a monk named Gregor Mendel performed experiments on heredity. His discoveries helped establish the field of genetics. *Genetics* is the study of how traits are passed on, or inherited. ☑

STUDY TIP

Define As you read this section, make a list of all of the underlined and italicized words. Write a definition for each of the words.

READING CHECK

1. Define What is genetics?

____

____

____

Who Was Gregor Mendel?

Gregor Mendel was born in Austria in 1822. He grew up on a farm where he learned a lot about flowers and fruit trees. When he was 21 years old, Mendel entered a monastery. A monastery is a place where monks study and practice religion. The monks at Mendel's monastery also taught science and performed scientific experiments.

Mendel studied pea plants in the monastery garden to learn how traits are passed from parents to offspring. He used garden peas because they grow quickly. They also have many traits, such as height and seed color, that are easy to see. His results changed the way people think about how traits are passed on. ☑

READING CHECK

2. Explain Why did Mendel choose to study pea plants?

____

____

____

____

Gregor Mendel discovered the principles of heredity while studying pea plants.

REPRODUCTION IN PEAS

Like many flowering plants, pea plants have both male and female reproductive parts. Many flowering plants can reproduce by cross-pollination. In most plants, sperm are carried in structures called pollen. In *cross-pollination*, sperm in the pollen of one plant fertilize eggs in the flower of another plant. Pollen can be carried by organisms, such as insects. It may also be carried by the wind from one flower to another.

Some flowering plants must use cross-pollination. They need another plant to reproduce. However, some plants, including pea plants, can also reproduce by self-pollination. In *self-pollination*, sperm from one plant fertilize the eggs of the same plant.

Mendel used self-pollination in pea plants to grow true-breeding plants for his experiments. When a *true-breeding* plant self-pollinates, its offspring all have the same traits as the parent. For example, a true-breeding plant with purple flowers always has offspring with purple flowers.

Critical Thinking

3. Compare What is the difference between cross-pollination and self-pollination?

____

____

____

____

____

____

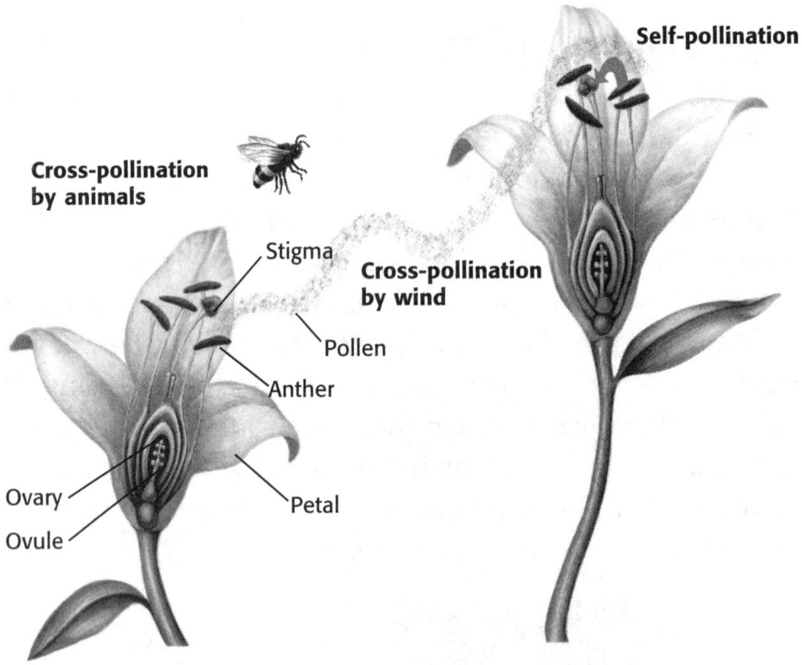
Self-pollination
Cross-pollination by animals
Stigma
Cross-pollination by wind
Pollen
Anther
Ovary
Ovule
Petal

During pollination, pollen from the anther (male) is carried to the stigma (female). Fertilization happens when a sperm from the pollen moves through the stigma and enters an egg in an ovule.

TAKE A LOOK

4. Identify What are two ways pollen can travel from one plant to another during cross-pollination?

____

____

____

____

CHARACTERISTICS

A *characteristic* is a feature that has different forms. For example, hair color is a characteristic of humans. The different forms or colors, such as brown or red hair, are *traits*. ☑

Mendel studied one characteristic of peas at a time. He used plants that had different traits for each characteristic he studied. One characteristic he studied was flower color. He chose plants that had purple flowers and plants that had white flowers. He also studied other characteristics, such as seed shape, pod color, and plant height.

READING CHECK

5. Explain How are characteristics and traits related?

____

____

____

____

CROSSING PEA PLANTS

Mendel was careful to use true-breeding plants in his experiments. By choosing these plants, he would know what to expect if his plants self-pollinated. He decided to find out what would happen if he bred, or crossed, two plants that had different traits.

Say It

Describe How would you describe yourself? Make a list of your physical traits, such as height, hair color, and eye color. List other traits you have that you weren't born with. Share this list with your classmates. Which of these traits did you inherit?

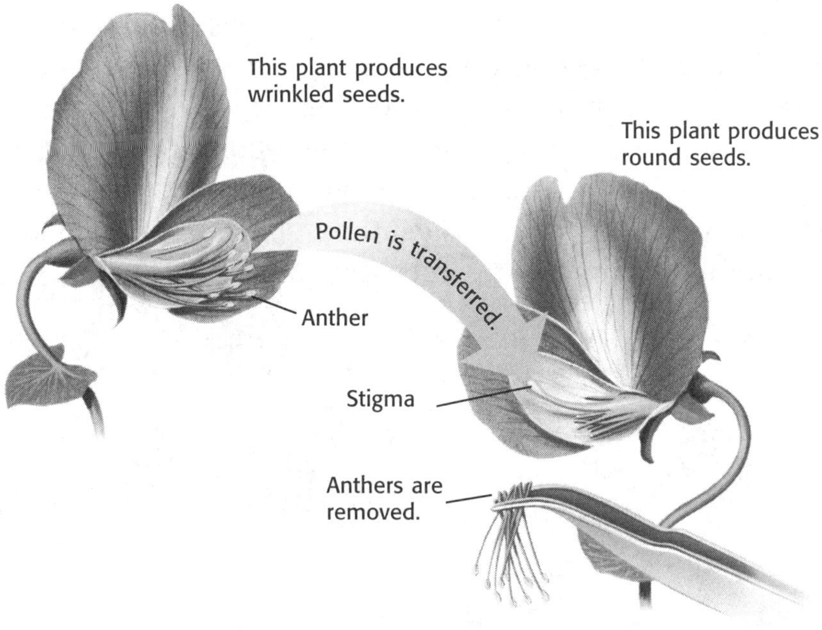
This plant produces wrinkled seeds.
This plant produces round seeds.
Pollen is transferred.
Anther
Stigma
Anthers are removed.

Mendel removed the anthers from a plant that made round seeds. Then, he used pollen from a plant that made wrinkled seeds to fertilize the plant that made round seeds.

TAKE A LOOK

6. Describe How did Mendel make sure that the plant with round seeds did not self-pollinate?

____

____

____

____

____

What Happened in Mendel's First Experiments?

Mendel studied seven different characteristics in his first experiments with peas. He crossed plants that were true-breeding for different traits. For example, he crossed plants that had purple flowers with plants that had white flowers. The offspring from such a cross are called *first-generation plants*. All of the first-generation plants in this cross had purple flowers. What happened to the trait for white flowers?

Mendel got similar results for each cross. One trait was always present in the first generation and the other trait seemed to disappear. Mendel called the trait that appeared the **dominant trait**. He called the other trait the **recessive trait**. To *recede* means "to go away or back off." To find out what happened to the recessive trait, Mendel did another set of experiments. ☑

READING CHECK

7. Identify What kind of trait appeared in the first generation?

____

What Happened in Mendel's Second Experiment?

Mendel let the first-generation plants self-pollinate. Some of the offspring were white-flowered, even though the parent was purple-flowered. The recessive trait for white flowers had reappeared in the second generation.

Mendel did the same experiment on plants with seven different characteristics. Each time, some of the second-generation plants had the recessive trait.

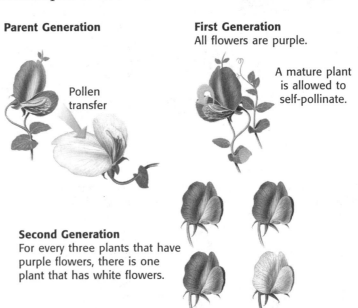

Parent Generation

First Generation
All flowers are purple.

Pollen
transfer
A mature plant
is allowed to
self-pollinate.

Second Generation
For every three plants that have purple flowers, there is one plant that has white flowers.

TAKE A LOOK

8. Identify What type of traits appeared in the second generation?

____

____

RATIOS IN MENDEL'S EXPERIMENTS

Mendel counted the number of plants that had each trait in the second generation. He hoped that this might help him explain his results.

As you can see from the table below, the recessive trait did not show up as often as the dominant trait. Mendel decided to figure out the ratio of dominant traits to recessive traits. A *ratio* is a relationship between two numbers. It is often written as a fraction. For example, the second generation produced 705 plants with purple flowers and 224 plants with white flowers. Mendel used this formula to calculate the ratios:

$$\frac{705}{224} = \frac{3.15}{1} \text{ or } 3.15:1$$

Characteristic	Dominant trait	Recessive trait	Ratio
Flower color	705 purple 	224 white 	3.15:1
Seed color	6,002 yellow 	2,001 green 	
Seed shape	5,474 round 	1,850 wrinkled 	
Pod color	428 green 	152 yellow 	
Pod shape	882 smooth 	299 bumpy 	
Flower position	651 along stem 	207 at tip 	
Plant height	787 tall 	277 short 	

Math Focus

9. Find Ratios Calculate the ratios of the other pea plant characteristics in the table.

Math Focus

10. Round Round off all numbers in the ratios to whole numbers. What ratio do you get?

____

What Did Mendel Conclude?

Mendel knew that his results could be explained only if each plant had two sets of instructions for each characteristic. He concluded that each parent gives one set of instructions to the offspring. The dominant set of instructions determines the offspring's traits.

Name ______________________ Class ______________ Date ______________

Section 1 Review

NSES LS 2b, 2e

SECTION VOCABULARY

dominant trait the trait observed in the first generation when parents that have different traits are bred

heredity the passing of genetic traits from parent to offspring

recessive trait a trait that is apparent only when two recessive alleles for the same characteristic are inherited

1. **Define** What is a true-breeding plant?

____

____

____

2. **Apply Concepts** Cats may have straight or curly ears. A curly-eared cat mated with a straight-eared cat. All the kittens had curly ears. Are curly ears a dominant or recessive trait? Explain your answer.

____

____

3. **Summarize** Complete the cause and effect map to summarize Mendel's experiments on flower color in pea plants.

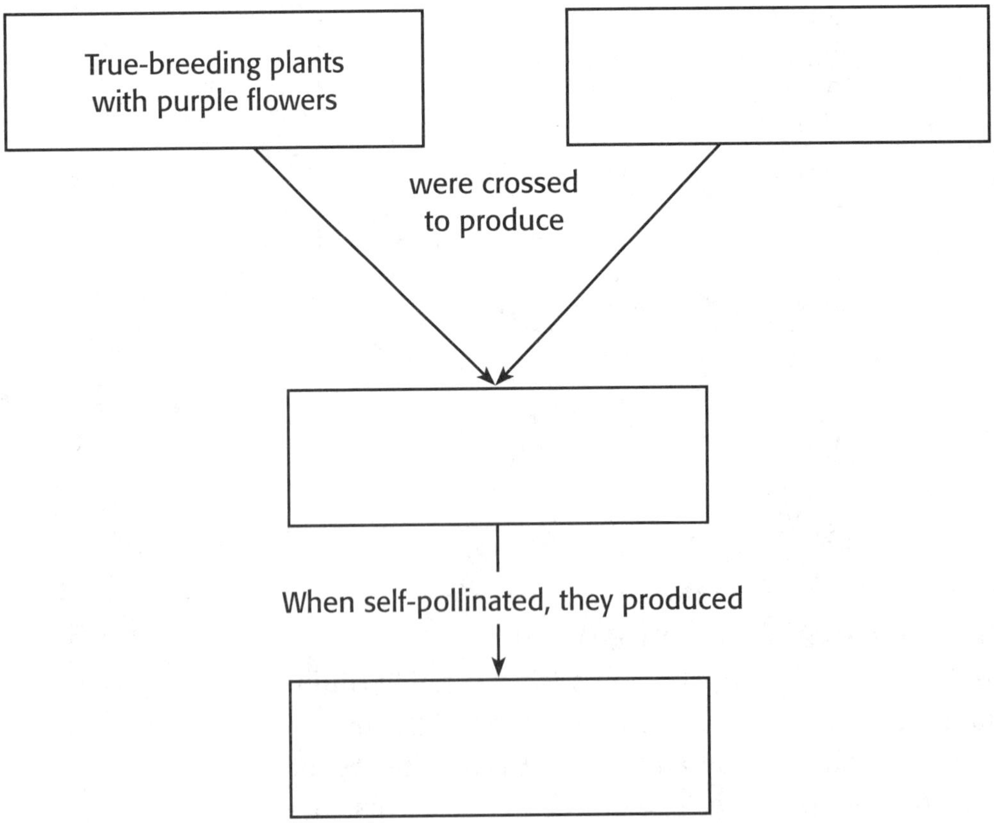
True-breeding plants with purple flowers
were crossed to produce
When self-pollinated, they produced

CHAPTER 3 Heredity

SECTION 2

Traits and Inheritance

BEFORE YOU READ

After you read this section, you should be able to answer these questions:

- What did Mendel's experiments tell him about heredity?
- Are there exceptions to Medel's laws of heredity?

National Science Education Standards
LS 2a, 2b, 2c, 2d, 2e

What Did Mendel Learn About Heredity?

Mendel knew from his pea plant experiments that there must be two sets of instructions for each characteristic. All of the first-generation plants showed the dominant trait. However, they could give the recessive trait to their offspring. Instructions for an inherited trait are called **genes**. Offspring have two sets of genes—one from each parent.

STUDY TIP
Organize As you read, make a Concept Map using the vocabulary words highlighted in the section.

The two sets of genes that parents give to offspring are never exactly the same. The same gene might have more than one version. The different versions of a gene are called **alleles**. ☑

READING CHECK
1. Define What is an allele?

____

____

____

Alleles may be dominant or recessive. A trait for an organism is usually identified with two letters, one for each allele. Dominant alleles are given capital letters (*A*). Recessive alleles are given lowercase letters (*a*). If a dominant allele is present, it will hide a recessive allele. An organism can have a recessive trait only if it gets a recessive allele for that trait from both parents.

PHENOTYPE

An organism's genes affect its traits. The appearance of an organism, or how it looks, is called its **phenotype**. The phenotypes for flower color in Mendel's pea plants were purple and white. The figure below shows one example of a human phenotype. ☑

READING CHECK
2. Define What is a phenotype?

____

____

Albinism is an inherited disorder that affects a person's phenotype in many ways.

GENOTYPE

A **genotype** is the combination of alleles that an oganism gets from its parents. A plant with two dominant or two recessive alleles (*PP*, *pp*) is *homozygous. Homo* means "the same." A plant with one dominant allele and one recessive allele (*Pp*) is *heterozygous. Hetero* means "different." The allele for purple flowers (*P*) in pea plants is dominant. The plant will have purple flowers even if it has only one *P* allele. ☑

READING CHECK

3. Identify What kind of alleles does a heterozygous individual have?

PUNNETT SQUARES

A Punnett square is used to predict the possible genotypes of offspring from certain parents. It can be used to show the alleles for any trait. In a Punnett square, the alleles for one parent are written along the top of the square. The alleles for the other parent are written along the side of the square. The possible genotypes of offspring are found by combining the letters at the top and side of each square.

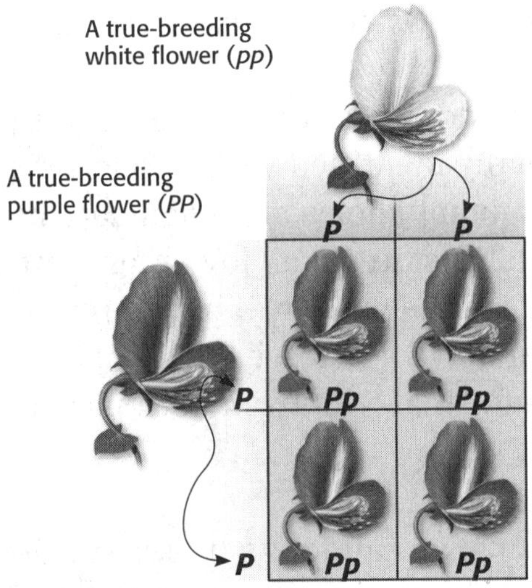
A true-breeding white flower (pp)
A true-breeding purple flower (PP)
p
p
P
P
Pp
Pp
Pp
Pp

All of the offspring for this cross have the same genotype—*Pp*.

TAKE A LOOK

4. Identify Is the plant with white flowers homozygous or heterozygous? How can you tell?

The figure shows a Punnett square for a cross of two true-breeding plants. One has purple flowers and the other has white flowers. The alleles for a true-breeding purple-flowered plant are written as *PP*. The alleles for a true-breeding white flowered plant are written as *pp*. Offspring get one of their two alleles from each parent. All of the offspring from this cross will have the same genotype: *Pp*. Because they have a dominant allele, all of the offspring will have purple flowers.

MORE EVIDENCE FOR INHERITANCE

In his second experiments, Mendel let the first-generation plants self-pollinate. He did this by covering the flowers of the plant. This way, no pollen from another plant could fertilize its eggs. The Punnett square below shows a cross of a plant that has the genotype *Pp*.

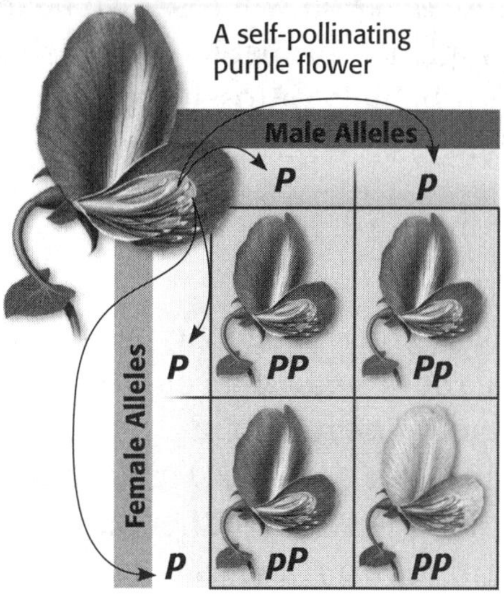
A self-pollinating purple flower
Male Alleles
P
p
Female Alleles
P
p
PP
Pp
pP
pp

This Punnett square shows the possible results from the cross *Pp* × *Pp*.

TAKE A LOOK

5. List What are the possible genotypes of the offspring in this cross?

____

____

Notice that one square shows the genotype *Pp* and another shows *pP*. These are exactly the same genotype. They both have one *p* allele and one *P* allele. The combinations *PP*, *Pp*, and *pP* have the same phenotype—purple flowers. This is because they all have at least one dominant allele, *P*. ☑

Only one combination, *pp*, produces plants that have white flowers. The ratio of dominant phenotypes to recessive phenotypes is 3:1. This means that three out of four offspring from that cross will have purple flowers. This is the same ratio Mendel found.

READING CHECK

6. Explain Why do the genotypes *PP*, *Pp*, and *pP* all have the same phenotype?

____

____

____

What Is the Chance That Offspring Will Receive a Certain Allele?

Each parent has two alleles for each gene. When an individual reproduces, it passes one of its two alleles to its offspring. When a parent has two different alleles for a gene, such as *Pp*, offspring may receive either of the alleles. Both alleles have an equal chance to be passed from the parent to the offspring.

Think of a coin toss. When you toss the coin, there is a 50% chance you will get heads, and a 50% chance you will get tails. The chance of the offspring receiving one allele or another from a parent is as random as a coin toss.

PROBABILITY

The mathematical chance that something will happen is known as **probability**. Probability is usually written as a fraction or percentage. If you toss a coin, the probability of tossing tails is 1/2, or 50%. In other words, you will get tails half of the time.

What is the probability that you will toss two heads in a row? To find out, multiply the probability of tossing the first head (1/2) by the probability of tossing the second head (1/2). The probability of tossing two heads in a row is 1/4.

GENOTYPE PROBABILITY

Finding the probability of certain genotypes for offspring is like predicting the results of a coin toss. To have white flowers, a pea plant must receive a *p* allele from each parent. Each offspring of a *Pp* × *Pp* cross has a 50% chance of receiving either allele from either parent. So, the probability of inheriting two *p* alleles is 1/2 × 1/2. This equals 1/4, or 25%.

Math Focus

7. Complete Complete the Punnett square to show the cross between two heterozygous parents. What percentage of the offspring are homozygous?

____

	P	*p*
P		
p		

Are There Exceptions to Mendel's Principles?

Mendel's experiments helped show the basic principles of how genes are passed from one generation to the next. Mendel studied sets of traits such as flower color and seed shape. The traits he studied in pea plants are easy to predict because there are only two choices for each trait. ☑

Traits in other organisms are often harder to predict. Some traits are affected by more than one gene. A single gene may affect more than one trait. As scientists learned more about heredity, they found exceptions to Mendel's principles.

READING CHECK

8. Explain Why were color and seed shape in pea plants good traits for Mendel to study?

____

____

____

INCOMPLETE DOMINANCE

Sometimes, one trait isn't completely dominant over another. These traits do not blend together, but each allele has an influence on the traits of offspring. This is called *incomplete dominance.* For example, the offspring of a true-breeding red snapdragon and a true-breeding white snapdragon are all pink. This is because both alleles for the gene influence color.

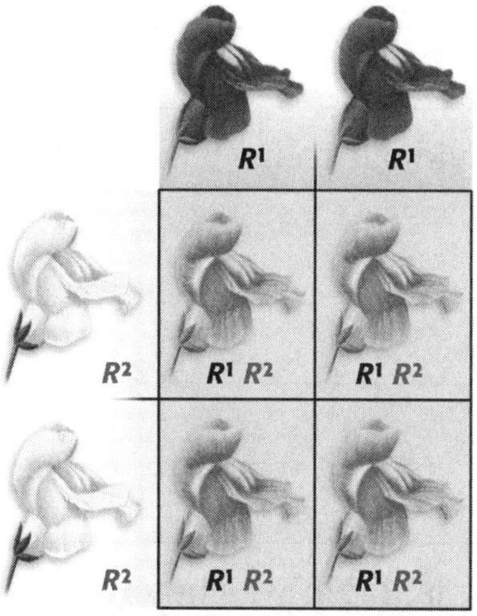
R¹
R¹
R²
R¹ R²
R¹ R²
R²
R¹ R²
R¹ R²

The offspring of two true-breeding show incomplete dominance.

ONE GENE, MANY TRAITS

In Mendel's studies, one gene controlled one trait. However, some genes affect more than one trait. For example, some tigers have white fur instead of orange. These white tigers also have blue eyes. This is because the gene that controls fur color also affects eye color.

MANY GENES, ONE TRAIT

Some traits, such as the color of your skin, hair, and eyes, are the result of several genes acting together. In humans, different combinations of many alleles can result in a variety of heights. ☑

THE IMPORTANCE OF ENVIRONMENT

Genes are not the only things that can affect an organism's traits. Traits are also affected by factors in the environment. For example, human height is affected not only by genes. Height is also influenced by nutrition. An individual who has plenty of food to eat may be taller than one who does not.

Critical Thinking

9. Infer If snapdragons showed complete dominance like pea plants, what would the offspring look like?

____

____

____

____

Critical Thinking

10. Compare How is the allele for fur color in tigers different from the allele for flower color in pea plants?

____

____

____

____

____

____

READING CHECK

11. Identify Give an example of a single trait that is affected by more than one gene.

____

Name ______________________ Class ______________ Date ______________

Section 2 Review

NSES LS 2a, 2b, 2c, 2d, 2e

SECTION VOCABULARY

allele one of the alternative forms of a gene that governs a characteristic, such as hair color

gene one set of instructions for an inherited trait

genotype the entire genetic makeup of an organism; also the combination of genes for one or more specific traits

phenotype an organism's appearance or other detectable characteristic

probability the likelihood that a possible future event will occur in any given instance of the event

1. **Identify Relationships** How are genes and alleles related?

____

____

2. **Explain** How is it possible for two individuals to have the same phenotype but different genotype for a trait?

____

____

____

____

3. **Punnett Square** Mendel allowed a pea plant that was heterozygous for yellow seeds (*Y*) to self-pollinate. Fill in the Punnett square below for this cross. What percentage of the offspring will have green (*y*) seeds?

____

	Y	*y*
Y		
y		

4. **Discuss** How is human height an exception to Mendel's principles of heredity?

____

____

____

Name ______________________ Class ______________ Date ______________

CHAPTER 3 Heredity

SECTION 3

Meiosis

BEFORE YOU READ

After you read this section, you should be able to answer these questions:

- What are sex cells?
- How does meiosis help explain Mendel's results?

National Science Education Standards

LS 1c, 1d, 2a, 2b, 2c, 2d

How Do Organisms Reproduce?

When organisms reproduce, their genetic information is passed on to their offspring. There are two kinds of reproduction: asexual and sexual.

ASEXUAL REPRODUCTION

In asexual reproduction, only one parent is needed to produce offspring. Asexual reproduction produces offspring with exact copies of the parent's genotype.

SEXUAL REPRODUCTION

In sexual reproduction, cells from two parents join to form offspring. Sexual reproduction produces offspring that share traits with both parents. However, the offspring are not exactly like either parent.

What Are Homologous Chromosomes?

Recall that genes are the instructions for inherited traits. Genes are located on chromosomes. Each human body cell has a total of 46 chromosomes, or 23 pairs. A pair of chromosomes that carry the same sets of genes are called **homologous chromosomes**. One chromosome from a pair comes from each parent. ☑

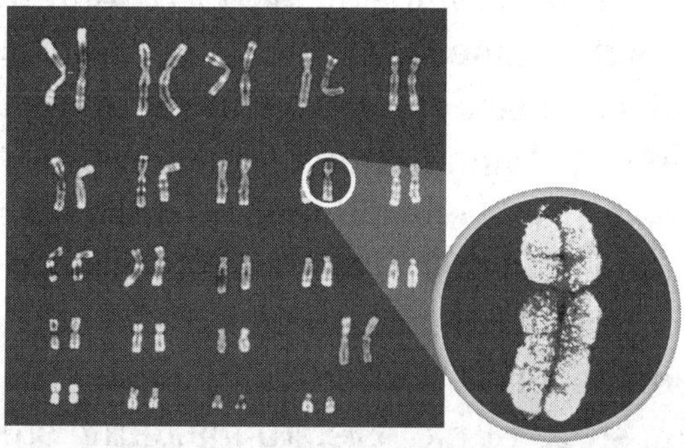

Human body cells have 23 pairs of chromosomes. One member of a pair of homologous chomosomes has been magnified.

STUDY TIP

Summarize Make flashcards that show the steps of meiosis. On the front of the cards, write the steps of meiosis. On the back of the cards, write what happens at each step. Practice arranging the steps in the correct order.

READING CHECK

1. Define What are homologous chromosomes?

____

____

____

TAKE A LOOK

2. Identify How many total chromosomes are in each human body cell?

____

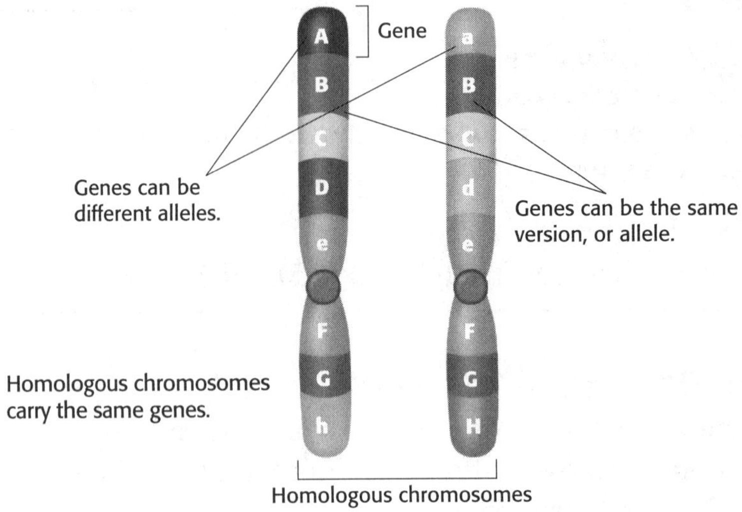
Gene
A
B
C
D
e
F
G
h
a
B
C
d
e
F
G
H
Genes can be different alleles.
Genes can be the same version, or allele.
Homologous chromosomes carry the same genes.
Homologous chromosomes

READING CHECK

3. Explain How are sex cells different from ordinary body cells?

____

____

____

STANDARDS CHECK

LS 2b In many species, including humans, females produce eggs and males produce sperm...An egg and sperm unite to begin development of a new individual. The individual receives genetic information from its mother (via the egg) and its father (via the sperm). Sexually produced offspring never are identical to either of their parents.

4. Define What is the function of meiosis?

____

____

____

____

What Are Sex Cells?

In sexual reproduction, cells from two parents join to make offspring. However, only certain cells can join. Cells that can join to make offspring are called *sex cells*. An egg is a female sex cell. A sperm is a male sex cell. Unlike ordinary body cells, sex cells do not have homologous chromosomes. ☑

Imagine a pair of shoes. Each shoe is like a chromosome and the pair represents a homologous pair of chromosomes. Recall that your body cells have a total of 23 pairs of "shoes," or homologous chromosomes. Each sex cell, however, has only one of the chromosomes from each homologous pair. Sex cells have only one "shoe" from each pair. How do sex cells end up with only one chromosome from each pair?

How Are Sex Cells Made?

Sex cells are made during meiosis. **Meiosis** is a copying process that produces cells with half the usual number of chromosomes. Meiosis keeps the total number of chromosomes the same from one generation to the next.

In meiosis, each sex cell that is made gets only one chromosome from each homologous pair. For example, a human egg cell has 23 chromosomes and a sperm cell has 23 chromosomes. When these sex cells later join together during reproduction, they form pairs. The new cell has 46 chromosomes, or 23 pairs. The figure on the next page describes the steps of meiosis. To make the steps easy to see, only four chromosomes are shown.

Steps of Meiosis

First cell division

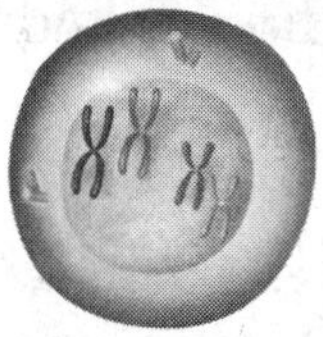

❶ The chromosomes are copied before meiosis begins. The identical copies, or chromatids, are joined together.

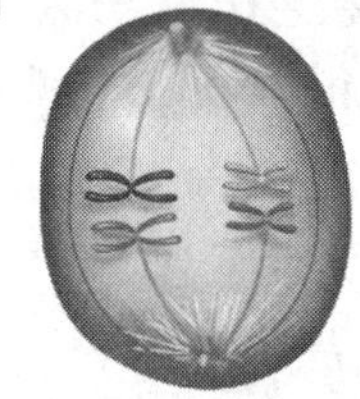

❷ The nuclear membrane disappears. Pairs of homologous chromosomes line up at the equator of the cell.

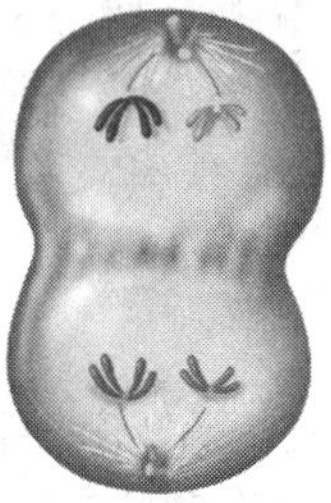

❸ The chromosomes separate from their homologous partners. Then they move to the opposite ends of the cell.

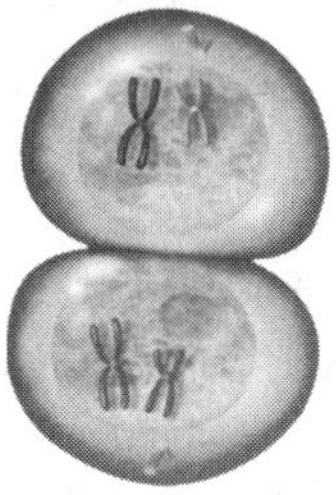

❹ The nuclear membrane re-forms, and the cell divides. The paired chromatids are still joined.

Second cell division

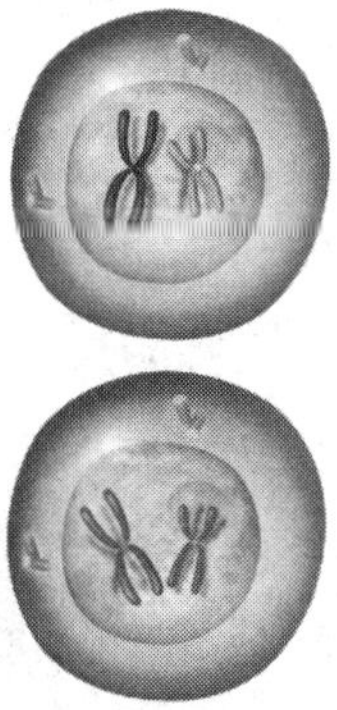

❺ Each cell contains one member of the homologous chromosome pair. The chromosomes are not copied again between the two cell divisions.

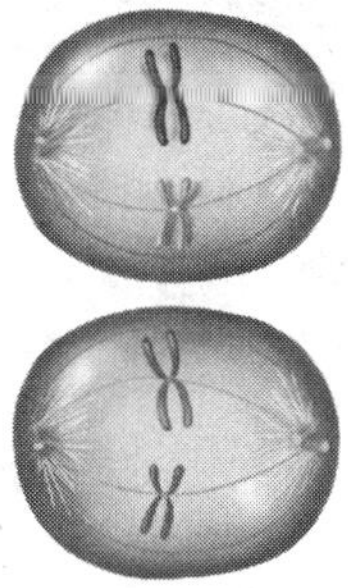

❻ The nuclear membrane disappears. The chromosomes line up along the equator of each cell.

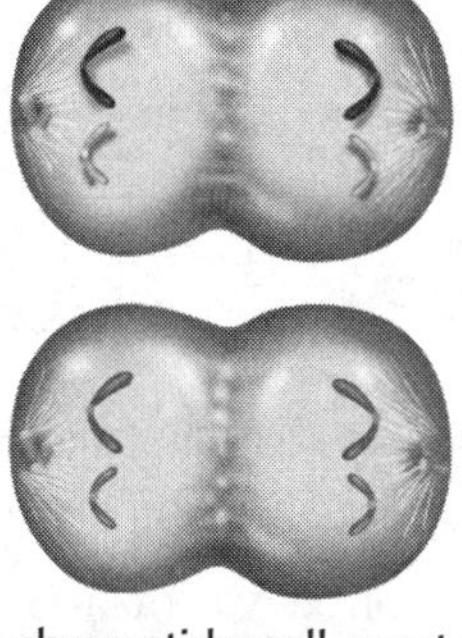

❼ The chromatids pull apart and move to opposite ends of the cell. The nuclear membranes re-form, and the cells divide.

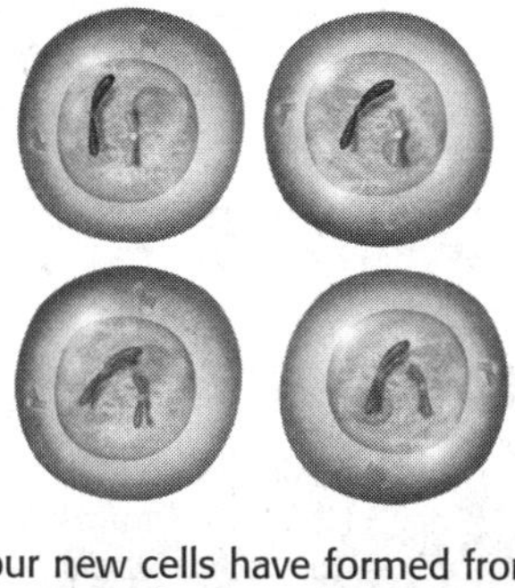

❽ Four new cells have formed from the original cell. Each new cell has half the number of chromosomes as the original cell.

Critical Thinking

5. Predict What would happen if meiosis did not occur?

____

____

____

____

____

TAKE A LOOK

6. Identify How many times does the cell nucleus divide during meiosis?

____

7. Identify At the end of meiosis, how many sex cells have been produced from one cell?

____

How Does Meiosis Explain Mendel's Results?

Mendel knew that eggs and sperm give the same amount of information to offspring. However, he did not know how traits were actually carried in the cell. Many years later, a scientist named Walter Sutton was studying grasshopper sperm cells. He knew about Mendel's work. When he saw chromosomes separating during meiosis, he made an important conclusion: genes are located on chromosomes.

The figure below shows what happens to chromosomes during meiosis and fertilization in pea plants. The cross shown is between two true-breeding plants. One produces round seeds and the other produces wrinkled seeds.

Critical Thinking

8. Identify Relationships How did Sutton's work build on Mendel's work?

____

____

____

____

____

____

Meiosis and Dominance

Male Parent In the plant cell nucleus below, each homologous chromosome has an allele for seed shape. Each allele carries the same instructions: to make wrinkled seeds.

Female Parent In the plant cell nucleus below, each homologous chromosome has an allele for seed shape. Each allele carries the same instructions: to make round seeds.

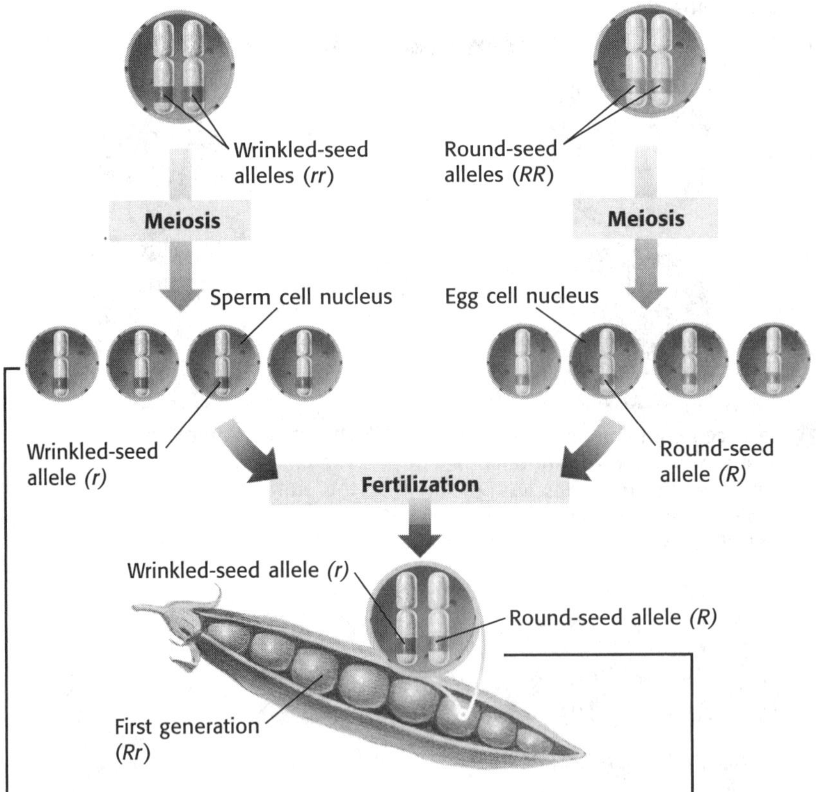
Wrinkled-seed alleles (*rr*)
Round-seed alleles (*RR*)
Meiosis
Meiosis
Sperm cell nucleus
Egg cell nucleus
Wrinkled-seed allele (*r*)
Round-seed allele (*R*)
Fertilization
Wrinkled-seed allele (*r*)
Round-seed allele (*R*)
First generation (*Rr*)

a After meiosis, each sperm cell has a recessive allele for wrinkled seeds. Each egg cell has a dominant allele for round seeds.

b Fertilization of any egg by any sperm gives the same genotype (*Rr*) and the same phenotype (round). This result is exactly what Mendel found in his studies.

TAKE A LOOK

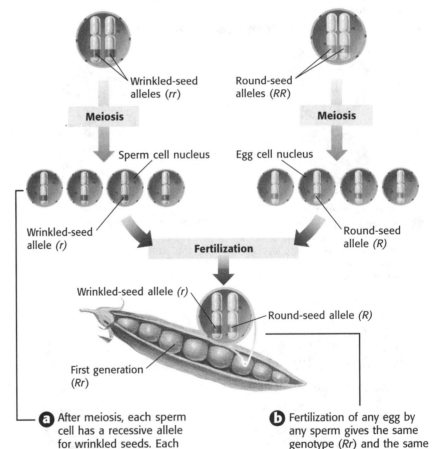

9. Explain In this figure, how many genotypes are possible for the offspring? Explain your answer.

____

____

____

____

____

What Are Sex Chromosomes?

Information contained on chromosomes determines many of our traits. **Sex chromosomes** carry genes that determine sex. In humans, females have two X chromosomes. Human males have one X chromosome and one Y chromosome. ☑

During meiosis, one of each of the chromosome pairs ends up in a sex cell. Females have two X chromosomes in each body cell. When meiosis produces egg cells, each egg gets one X chromosome. Males have both an X chromosome and a Y chromosome in each body cell. Meiosis produces sperm with either an X or a Y chromosome.

An egg fertilized by a sperm with an X chromosome will produce a female. If the sperm contains a Y chromosome, the offspring will be male.

READING CHECK

10. Identify What combination of sex chromosomes makes a human male?

____

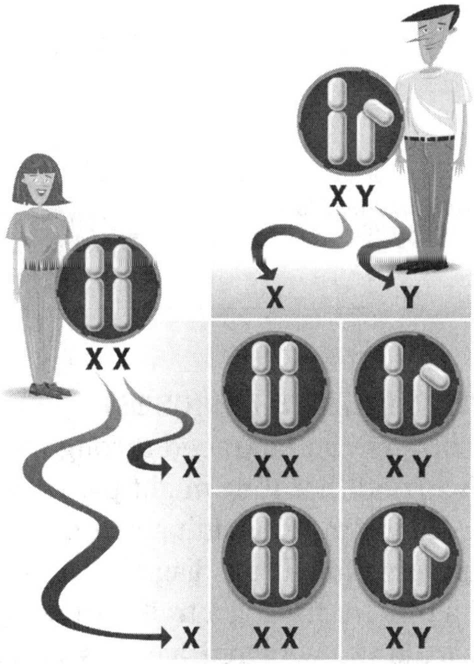
XY
X
Y
XX
X
XX
XY
X
XX
XY

Egg and sperm join to form either the XX or XY combination.

TAKE A LOOK

11. Identify Circle the offspring in the figure that will be female.

SEX-LINKED DISORDERS

Hemophilia is a disorder that prevents blood from clotting. People with hemophilia bleed for a long time after small cuts. This disorder can be fatal. Hemophilia is an example of a sex-linked disorder. The genes for *sex-linked disorders* are carried on the X chromosome. Colorblindness is another example of a sex-linked disorder. Men are more likely than women to have sex-linked disorders. Why is this? ☑

READING CHECK

12. Define What is a sex-linked disorder?

____

____

____

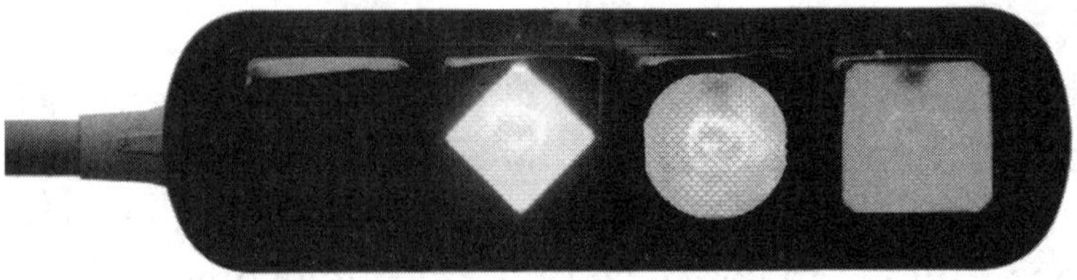

This stoplight in Canada was made to help the colorblind see signals easily.

The Y chromosome does not carry all of the genes that an X chromosome does. Females have two X chromosomes, so they carry two copies of each gene found on the X chromosome. This makes a backup gene available if one becomes damaged. Males have only one copy of each gene on their one X chromosome. If a male gets an allele for a sex-linked disorder, he will have the disorder, even if the allele is recessive.

TAKE A LOOK

13. Complete A particular sex-linked disorder is recessive. Fill in the Punnett Square to show how the disorder is passed from a carrier to its offspring. The chromosome carrying the trait for the disorder is underlined.

14. Identify Which individual will have the disorder?

____

	X	***Y***
$\underline{X}$		
X		

GENETIC COUNSELING AND PEDIGREES

Genetic disorders can be traced through a family tree. If people are worried that they might pass a disease to their children, they may consult a genetic counselor.

These counselors often use a diagram called a pedigree. A **pedigree** is a tool for tracing a trait through generations of a family. By making a pedigree, a counselor can often predict whether a person is a carrier of a hereditary disease.

The pedigree on the next page traces a disease called *cystic fibrosis*. Cystic fibrosis causes serious lung problems. People with this disease have inherited two recessive alleles. Both parents need to be carriers of the gene for the disease to show up in their children.

Pedigree for a Recessive Disease

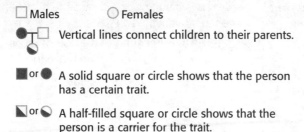

☐ Males ◯ Females

Vertical lines connect children to their parents.

■ or ● A solid square or circle shows that the person has a certain trait.

◪ or ◐ A half-filled square or circle shows that the person is a carrier for the trait.

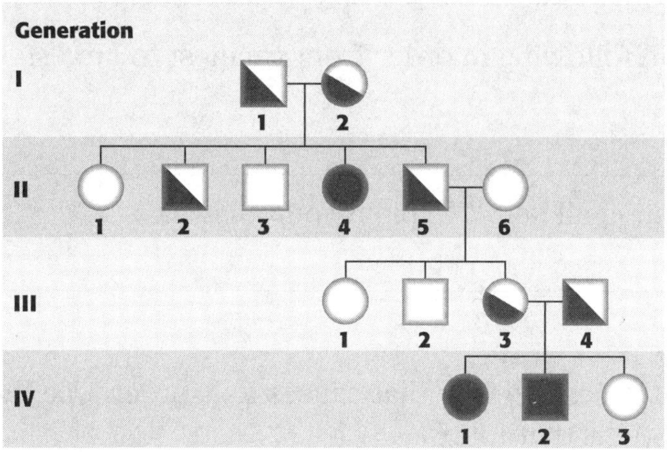
Generation
I
1 2
II
1 2 3 4 5 6
III
1 2 3 4
IV
1 2 3

TAKE A LOOK

15. Identify Circle all of the individuals in the pedigree who have the disorder. Draw a line under the individuals that carry the trait, but do not have the disorder.

You could draw a pedigree to trace almost any trait through a group of people who are biologically related. For example, a pedigree can show how you inherited your hair color. Many different pedigrees could be drawn for related individuals.

What Is Selective Breeding?

For thousands of years, humans have bred plants and animals to produce individuals with traits that they liked. This is known as *selective breeding*. Breeders may choose a plant or animal with traits they would like to see in the offspring. They breed that individual with another that also has those traits. For example, farmers might breed fruit trees that bear larger fruits.

You may see example of selective breeding every day. Different breeds of dogs, such as chihuahuas and German sheperds, were produced by selective breeding. Many flowers, such as roses, have been bred to produce large flowers. Wild roses are usually much smaller than roses you would buy at a flower store or plant nursery.

Say It

Discuss In a small group, come up with other examples of organisms that humans have changed through selective breeding. What traits do you think people wanted the organism to have? How is this trait helpful to humans?

Name ________________ Class ________________ Date ________________

Section 3 Review

NSES **LS 1c, 1d, 2a, 2b, 2c, 2d**

SECTION VOCABULARY

homologous chromosomes chromosomes that have the same sequence of genes and the same structure

meiosis a process in cell division during which the number of chromosomes decreases to half the original number by two divisions of the nucleus, which results in the production of sex cells (gametes or spores)

pedigree a diagram that shows the occurrence of a genetic trait in several generations of a family

sex chromosomes one of the pair of chromosomes that determine the sex of an individual

1. Identify Relationships Put the following in order from smallest to largest: chromosome, gene, cell.

____

2. Explain Does meiosis happen in all cells? Explain your answer.

____

The pedigree below shows a recessive trait that causes a disorder. Use the pedigree to answer the questions that follow.

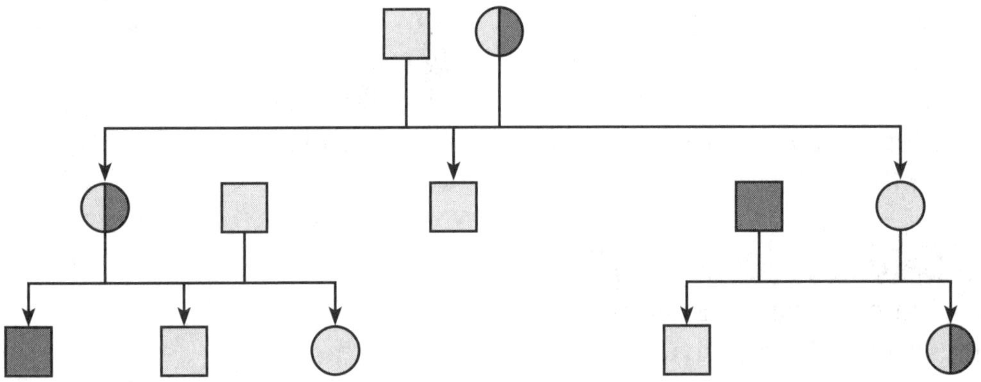

3. Identify Circle all individuals on the pedigree that are heterozygous for the trait. Are these individuals male or female?

____

4. Identify Put a square around all individuals that have the disorder. Are these individuals male or female?

____

5. Interpret Is the trait sex-linked? Explain your answer.

____

____

SECTION 1

What Does DNA Look Like?

BEFORE YOU READ

After you read this section, you should be able to answer these questions:

- What units make up DNA?
- What does DNA look like?
- How does DNA copy itself?

National Science Education Standards
LS 1a, 2d, 5a

What Is DNA?

Remember that *inherited traits* are traits that are passed from generation to generation. To understand how inherited traits are passed on, you must understand the structure of DNA. **DNA** (***d****eoxyribo****n****ucleic* ***a****cid*) is the molecule that carries the instructions for inherited traits. In cells, DNA is wrapped around proteins to form *chromosomes*. Stretches of DNA that carry the information for inherited traits are called *genes*.

STUDY TIP
Clarify Concepts As you read the text, make a list of ideas that are confusing. Discuss these with a small group. Ask your teacher to explain things that your group is unsure about.

What Is DNA Made Of?

DNA is made up of smaller units called nucleotides. A *nucleotide* is made of three parts: a sugar, a phosphate, and a base. The sugar and the phosphate are the same for each nucleotide. However, different nucleotides may have different bases.

There are four different bases found in DNA nucleotides. They are *adenine*, *thymine*, *guanine*, and *cytosine*. Scientists often refer to a base by its first letter: *A* for adenine, *T* for thymine, *G* for guanine, and *C* for cytosine. Each base has a different shape.

The Four Nucleotides of DNA

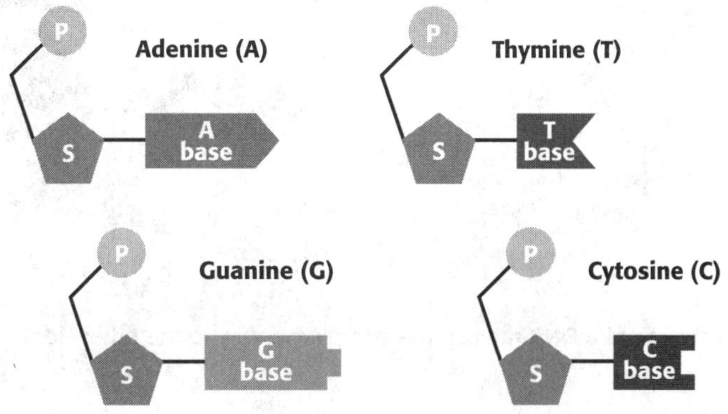
P
Adenine (A)
S
A base
P
Thymine (T)
S
T base
P
Guanine (G)
S
G base
P
Cytosine (C)
S
C base

TAKE A LOOK
1. Identify What are two things that are the same in all nucleotides?

____

____

What Does DNA Look Like?

As you can see in the figure below, a strand of DNA looks like a twisted ladder. This spiral shape is called a *double helix*. The two sides of the ladder are made of the sugar and phosphate parts of nucleotides. The sugars and phosphates alternate along each side of the ladder. The rungs of the DNA ladder are made of pairs of bases. ☑

READING CHECK

2. Identify What are the sides of the DNA "ladder" made of?

______

______

The bases in DNA can only fit together in certain ways, like puzzle pieces. Adenine on one side of a DNA strand always pairs with thymine on the other side. Guanine always pairs with cytosine. This means that adenine is *complementary* to thymine, and guanine is complementary to cytosine. Because the pairs of bases in DNA are complementary, the two sides of a strand of DNA are also complementary.

Critical Thinking

3. Apply Concepts Imagine that you are a scientist studying DNA. You measure the number of cytosines and thymines in a small strand of DNA. There are 45 cytosines and 55 thymines. How many guanines are there in the strand? How many adenines are there?

______

______

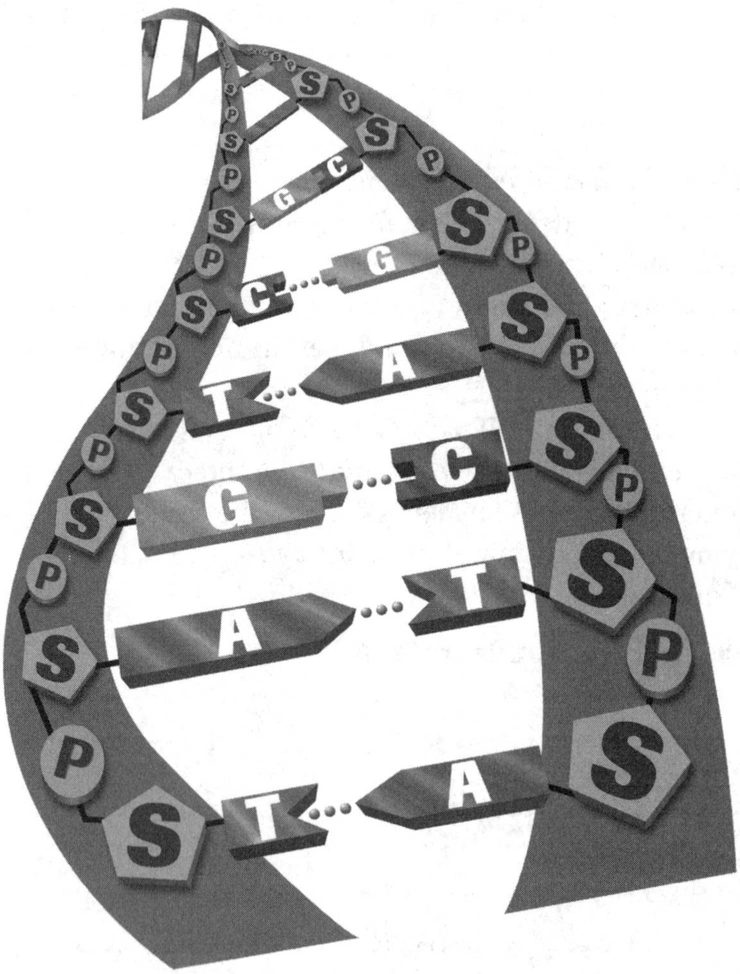
S
P
G
C
T
A

Each side of a DNA molecule is complementary to the other side.

TAKE A LOOK

4. Identify Give the ways that DNA bases can pair up.

______

______

______

How Does DNA Copy Itself?

Before a cell divides, it makes a copy of its genetic information for the new cell. The pairing of bases allows the cell to *replicate*, or make copies of, DNA. Remember that bases are complementary and can only fit together in certain ways. Therefore, the order of bases on one side of the DNA strand controls the order of bases on the other side of the strand. For example, the base order CGAC can only fit with the order GCTG. ☑

When DNA replicates, the pairs of bases separate and the DNA splits into two strands. The bases on each side of the original strand are used as a pattern to build a new strand. As the bases on the original strands are exposed, the cell adds nucleotides to form a new strand.

Finally, two DNA strands are formed. Half of each of the two DNA strands comes from the original strand. The other half is built from new nucleotides.

READING CHECK

5. Explain What happens to DNA before a cell divides?

____

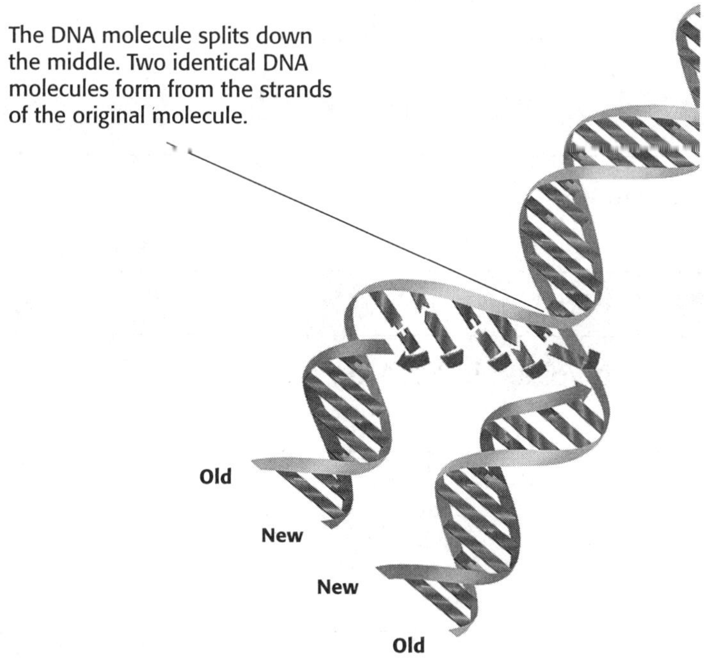
The DNA molecule splits down the middle. Two identical DNA molecules form from the strands of the original molecule.
Old
New
New
Old

TAKE A LOOK

6. Compare What is the difference between an "old" and a "new" strand of DNA?

____

____

____

DNA is copied every time a cell divides. Each new cell gets a complete copy of the entire DNA strand. Proteins in the cell unwind, copy, and rewind the DNA.

Name ______________________ Class ______________ Date ______________

Section 1 Review

NSES LS 1a, 2d, 5a

SECTION VOCABULARY

DNA deoxyribonucleic acid, a molecule that is present in all living cells and that contains the information that determines the traits that a living thing inherits and needs to live	**nucleotide** in a nucleic-acid chain, a subunit that consists of a sugar, a phosphate, and a nitrogenous base

1. Identify Where are genes located? What do they do?

____

____

2. Compare How are the four kinds of DNA nucleotides different from each other?

____

3. Apply Concepts The diagram shows part of a strand of DNA. Using the order of bases given in the top of the strand, write the letters of the bases that belong on the bottom strand.

C
T
T
A
G
G
C
T
T
A
C
C
A

____

4. Describe How is DNA related to chromosomes?

____

____

5. Identify Relationships How are proteins involved in DNA replication?

____

____

6. List What are three parts of a nucleotide?

____

____

____

Name ______________________ Class ______________ Date ______________

CHAPTER 4 Genes and DNA

SECTION 2

How DNA Works

BEFORE YOU READ

After you read this section, you should be able to answer these questions:

- What does DNA look like in different cells?
- How does DNA help make proteins?
- What happens if a gene changes?

National Science Education Standards
LS 1c, 1e, 1f, 2b, 2c, 2d, 2e, 5c

What Does DNA in Cells Look Like?

The human body contains trillions of cells, which carry out many different functions. Most cells are very small and can only be seen with a microscope. A typical skin cell, for example, has a diameter of about 0.0025 cm. However, almost every cell contains about 2 m of DNA. How can so much DNA fit into the nucleus of such a small cell? The DNA is bundled.

STUDY TIP

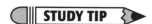

Compare After you read this section, make a table comparing chromatin, chromatids, and chromosomes.

Math Focus

1. Convert About how long is the DNA in a cell in inches?

1 in. = 2.54 cm

____

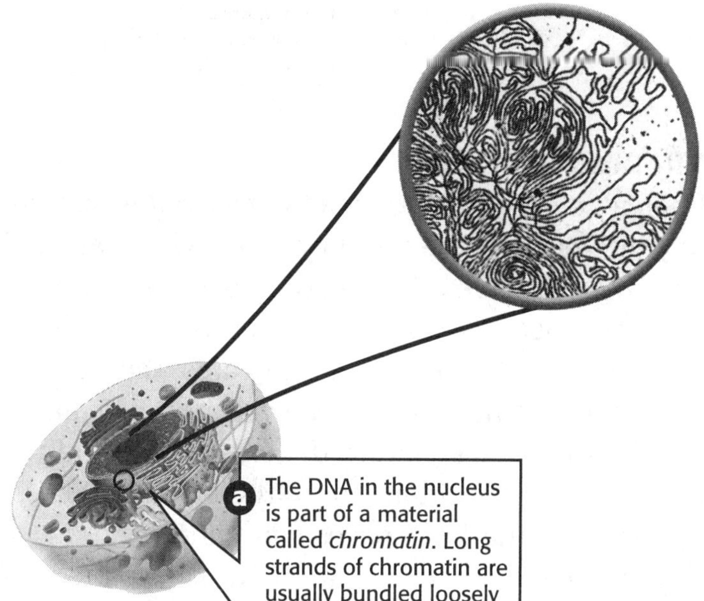

a The DNA in the nucleus is part of a material called *chromatin*. Long strands of chromatin are usually bundled loosely inside the nucleus.

TAKE A LOOK

2. Identify In what form is the DNA in the nucleus?

____

FITTING DNA INTO THE CELL

Large amounts of DNA can fit inside a cell because the DNA is tightly bundled by proteins. The proteins found with DNA help support the structure and function of DNA. Together, the DNA and the proteins it winds around make up a chromosome. ☑

DNA's structure allows it to hold a lot of information. Remember that a gene is made of a string of nucleotides. That is, it is part of the 2 m of DNA in a cell. Because there is an enormous amount of DNA, there can be a large variety of genes.

READING CHECK

3. Identify What are two things that are found in a chromosome?

____

____

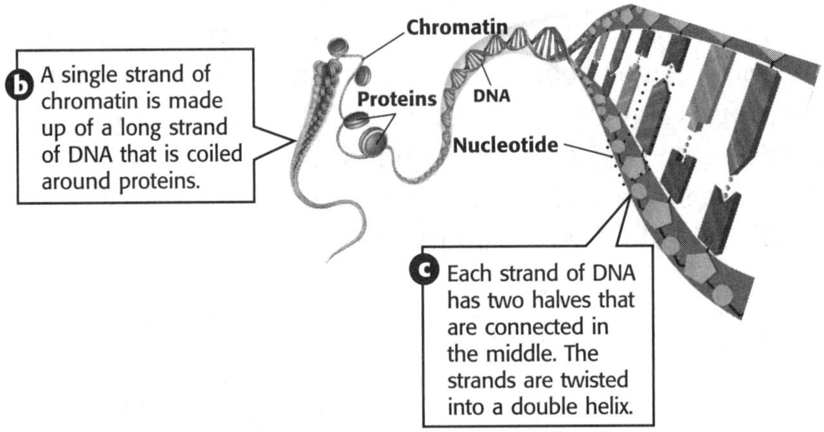
b A single strand of chromatin is made up of a long strand of DNA that is coiled around proteins.
Chromatin
Proteins
DNA
Nucleotide
c Each strand of DNA has two halves that are connected in the middle. The strands are twisted into a double helix.

TAKE A LOOK

4. Describe What is chromatin made of?

____

____

DNA IN DIVIDING CELLS

When a cell divides, its genetic material is spread equally into each of the two new cells. How can each of the new cells receive a full set of genetic material? It is possible because DNA replicates before a cell divides.

Remember that when DNA replicates, the strand of DNA splits down the middle. New strands are made when free nucleotide bases bind to the exposed strands. Each of the new strands is identical to the original DNA strand. This is because the DNA bases can join only in certain ways. *A* always pairs with *T*, and *C* always pairs with *G*.

Critical Thinking

5. Predict Consequences Imagine that DNA did not replicate before cell division. What would happen to the amount of DNA in each of the new cells formed during cell division?

____

____

____

____

____

When a cell is ready to divide, it has already copied its DNA. The copies stay attached as two chromatids. The two identical chromatids form a chromosome.

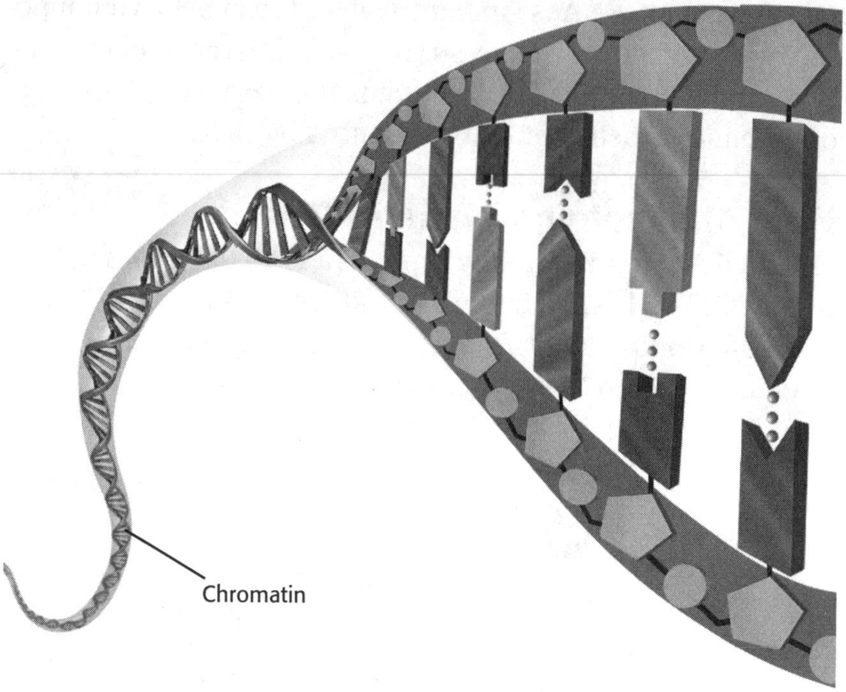
Chromatin

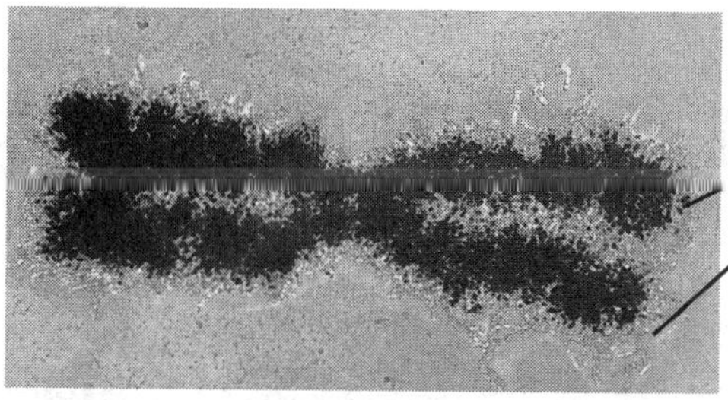
Chromatids

STANDARDS CHECK

LS 2d Hereditary information is contained in the genes, located in the chromosomes of each cell. Each gene carries a single unit of information. An inherited trait of an individual can be determined by one or by many genes, and a single gene can influence more than one trait. A human cell contains many thousands of different genes.

6. Identify Where is DNA found in a eukaryotic cell?

____

How Does DNA Help Make Proteins?

Proteins are found throughout cells. They cause most of the differences that you can see among organisms. A single organism can have thousands of different proteins.

Proteins act as chemical messengers for many of the activities in cells, helping the cells to work together. They also affect traits, such as the color of your eyes and how tall you will grow.

Proteins are made from many subunits called *amino acids*. A long string of amino acids forms a protein.

The order of bases in DNA is a code. The code tells how to make proteins. A group of three DNA bases acts as a code for one amino acid. For example, the group of DNA bases CAA *codes for*, or stands for, the amino acid valine. A gene usually contains instructions for making one specific protein.

Math Focus

7. Calculate How many DNA bases are needed to code for five amino acids?

____

HELP FROM RNA

RNA, or ***ribonucleic acid***, is a chemical that helps DNA make proteins. RNA is similar to DNA. It can act as a temporary copy of part of a DNA strand. One difference between DNA and RNA is that RNA contains the base *uracil* instead of thymine. Uracil is often represented by *U.* ☑

READING CHECK

8. Identify What is one difference between RNA and DNA?

____

____

____

How Are Proteins Made in Cells?

The first step in making a protein is to copy one side of part of the DNA. This mirrorlike copy is made of RNA. It is called *messenger RNA* (mRNA). It moves out of the nucleus and into the cytoplasm of the cell.

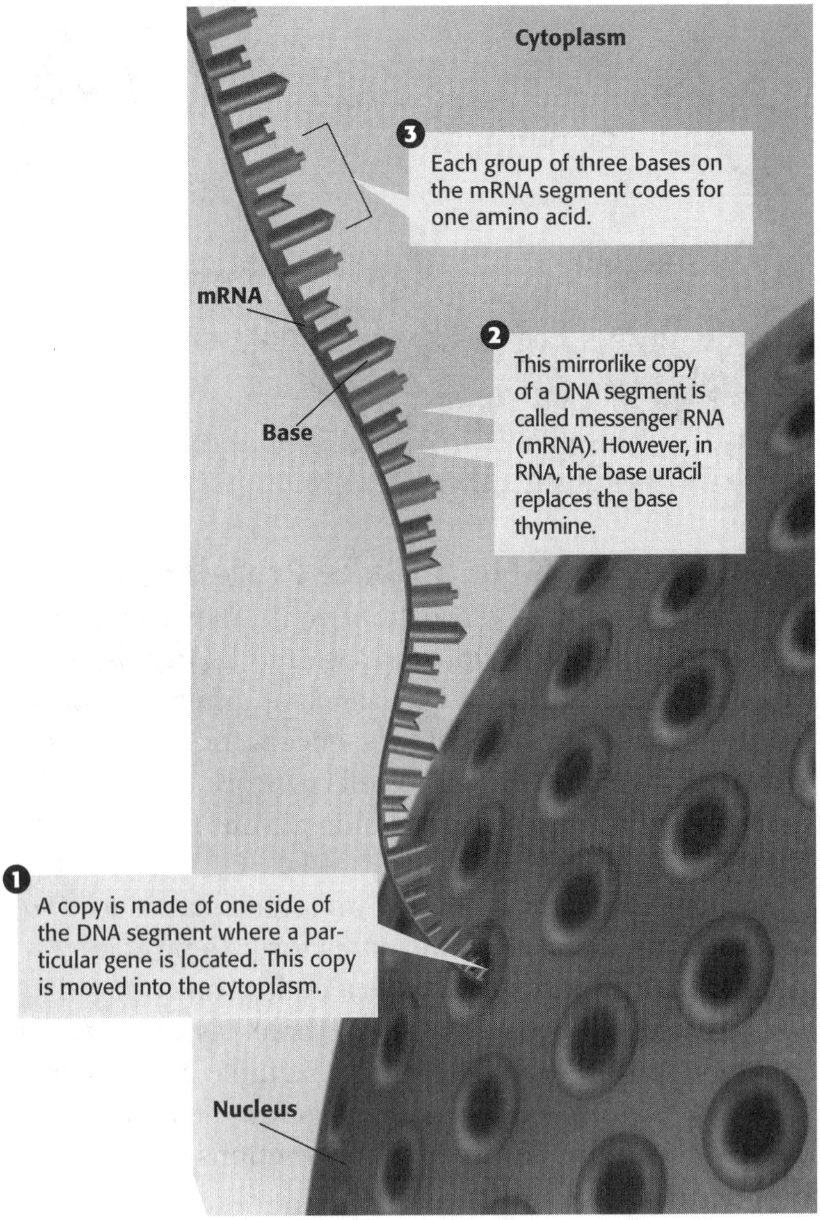
Cytoplasm
3 Each group of three bases on the mRNA segment codes for one amino acid.
mRNA
2 This mirrorlike copy of a DNA segment is called messenger RNA (mRNA). However, in RNA, the base uracil replaces the base thymine.
Base
1 A copy is made of one side of the DNA segment where a particular gene is located. This copy is moved into the cytoplasm.
Nucleus

TAKE A LOOK

9. Compare How does the shape of RNA differ from the shape of DNA?

____

____

____

RIBOSOMES

In the cytoplasm, the messenger RNA enters a protein assembly line. The "factory" that runs this assembly line is a ribosome. A **ribosome** is a cell organelle composed of RNA and protein. The mRNA moves through a ribosome as a protein is made.

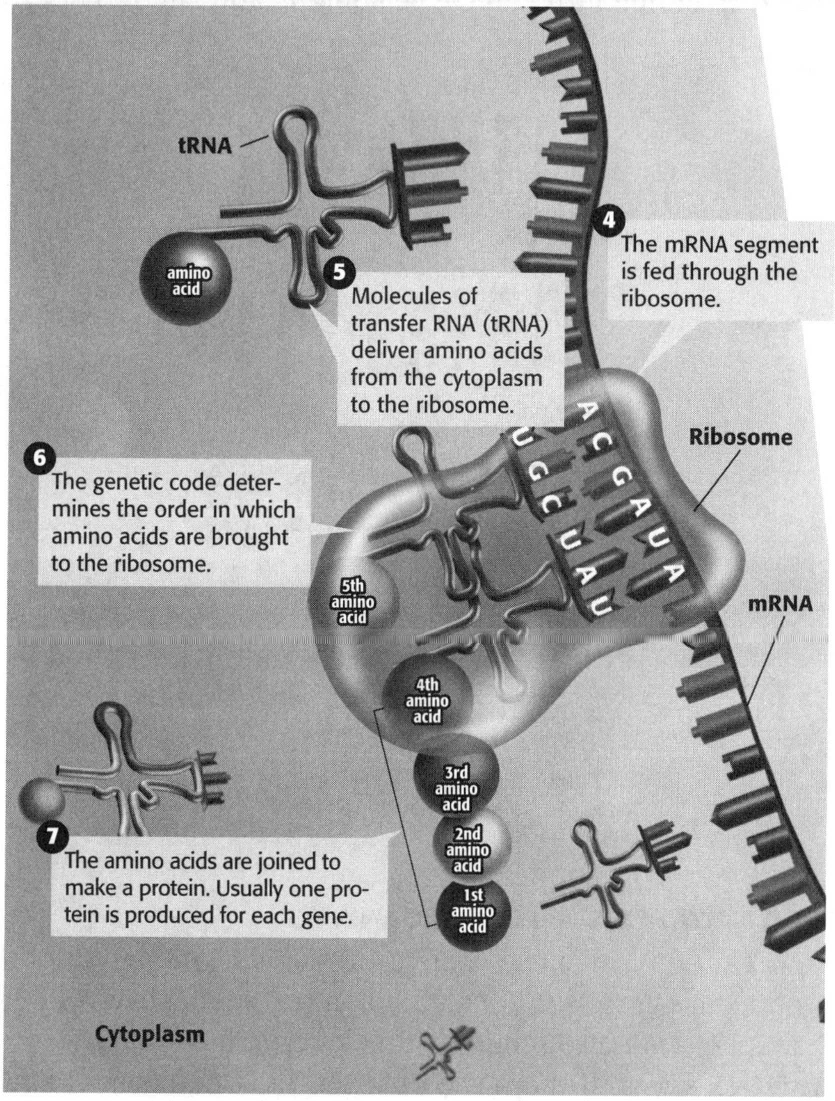
tRNA
amino acid
4 The mRNA segment is fed through the ribosome.
5 Molecules of transfer RNA (tRNA) deliver amino acids from the cytoplasm to the ribosome.
6 The genetic code determines the order in which amino acids are brought to the ribosome.
Ribosome
mRNA
5th amino acid
4th amino acid
3rd amino acid
2nd amino acid
1st amino acid
7 The amino acids are joined to make a protein. Usually one protein is produced for each gene.
Cytoplasm

Critical Thinking

10. Explain Proteins are made in the cytoplasm, but DNA never leaves the nucleus of a cell. How does DNA control how proteins are made?

____

____

____

____

____

____

____

TAKE A LOOK

11. Identify What does tRNA do?

____

____

____

What Happens If Genes Change?

Read this sentence: "Put the book on the desk." Does it make sense? What about this sentence: "Rut the zook in the tesk."? Changing only a few letters in a sentence can change what the sentence means. It can even keep the sentence from making any sense at all! In a similar way, even small changes in a DNA sequence can affect the protein that the DNA codes for. A change in the nucleotide-base sequence of DNA is called a **mutation**. ☑

☑ READING CHECK

12. Define What is a mutation?

____

____

____

HOW MUTATIONS HAPPEN

Some mutations happen because of mistakes when DNA is copied. Other mutations happen when DNA is damaged. Things that can cause mutations are called *mutagens*. Examples of mutagens include X rays and ultraviolet radiation. Ultraviolet radiation is one type of energy in sunlight. It can cause suntans and sunburns.

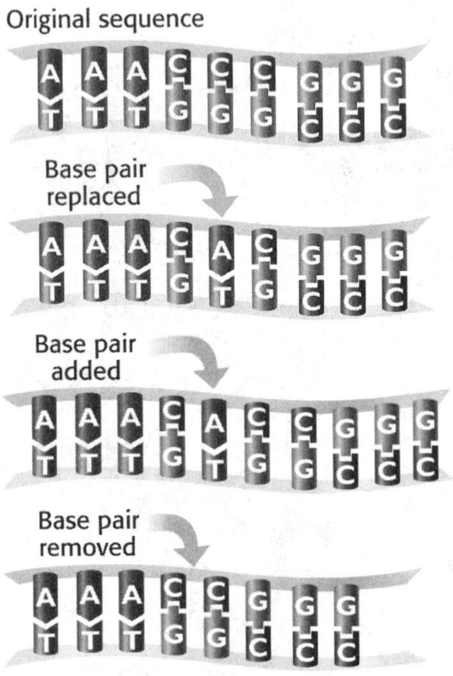
Original sequence
A A A C C C G G G
T T T G G G C C C
Base pair replaced
A A A C A C G G G
T T T G T G C C C
Base pair added
A A A C A C C G G G
T T T G T G G C C C
Base pair removed
A A A C C G G G
T T T G G C C C

Mutations can happen in different ways. A nucleotide may be replaced, added, or removed.

TAKE A LOOK

13. Compare What happens to one strand of DNA when there is a change in a base on the other strand?

____

____

HOW MUTATIONS AFFECT ORGANISMS

Mutations can cause changes in traits. Some mutations produce new traits that can help an organism survive. For example, a mutation might allow an organism to survive with less water. If there is a drought, the organism will be more likely to survive.

Many mutations produce traits that make an organism less likely to survive. For example, a mutation might make an animal a brighter color. This might make the animal easier for predators to find.

Some mutations are neither helpful nor harmful. If a mutation does not cause a change in a protein, then the mutation will not help or hurt the organism.

Say It

Brainstorm Whether a mutation is helpful or harmful to an organism often depends on the organism's environment. In a group, discuss how the same mutation could be helpful in one environment but harmful in another.

PASSING ON MUTATIONS

Cells make proteins that can find and fix many mutations. However, not all mutations can be fixed.

If a mutation happens in egg or sperm cells, the changed gene can be passed from one generation to the next. For example, sickle cell disease is caused by a genetic mutation that can be passed to future generations.

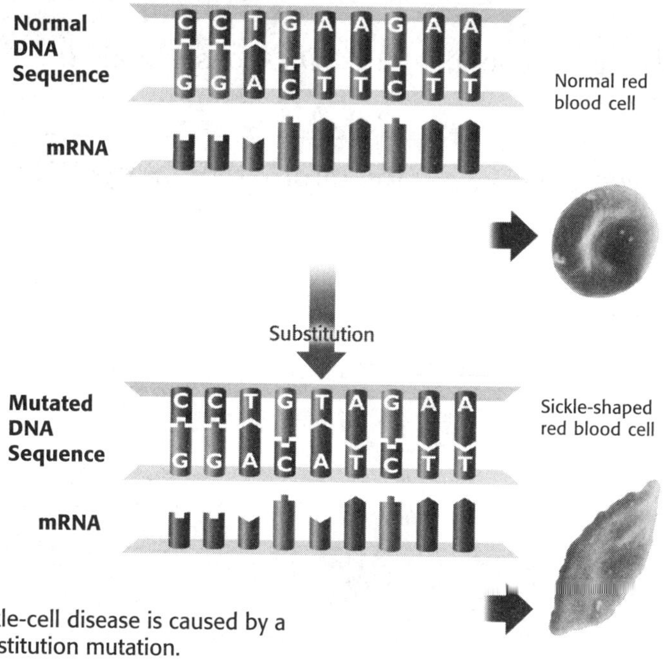
Normal DNA Sequence
C C T G A A G A A
G G A C T T C T T
mRNA
Normal red blood cell
Substitution
Mutated DNA Sequence
C C T G T A G A A
G G A C A T C T T
mRNA
Sickle-shaped red blood cell

Sickle-cell disease is caused by a substitution mutation.

TAKE A LOOK

14. Identify What kind of mutation causes sickle cell disease: deletion, insertion or substitution?

____

How Can We Use Genetic Knowledge?

Scientists use their knowledge of genetics in many ways. Most of these ways are helpful to people. However, other ways can cause ethical and scientific concerns.

GENETIC ENGINEERING

Scientists have learned how to change individual genes within organisms. This is called *genetic engineering*. In some cases, scientists transfer genes from one organism to another. For example, scientists can transfer genes from people into bacteria. The bacteria can then make proteins for people who are sick. ☑

READING CHECK

15. Define What is genetic engineering?

____

____

GENETIC IDENTIFICATION

Your DNA is unique, so it can be used like a fingerprint to identify you. *DNA fingerprinting* identifies the unique patterns in a person's DNA. Scientists can use these genetic fingerprints as evidence in criminal cases. They can also use genetic information to determine whether people are related.

Name ______________________ Class ______________ Date ______________

Section 2 Review

NSES LS 1c, 1e, 1f, 2b, 2c, 2d, 2e, 5c

SECTION VOCABULARY

mutation a change in the nucleotide-base sequence of a gene or DNA molecule

ribosome a cell organelle composed of RNA and protein; the site of protein synthesis

RNA ribonucleic acid, a molecule that is present in all living cells and that plays a role in protein production

1. Identify What structures in cells contain DNA and proteins?

____

2. Calculate How many amino acids can a sequence of 24 DNA bases code for?

____

3. Explain Fill in the flow chart below to show how the information in the DNA code becomes a protein.

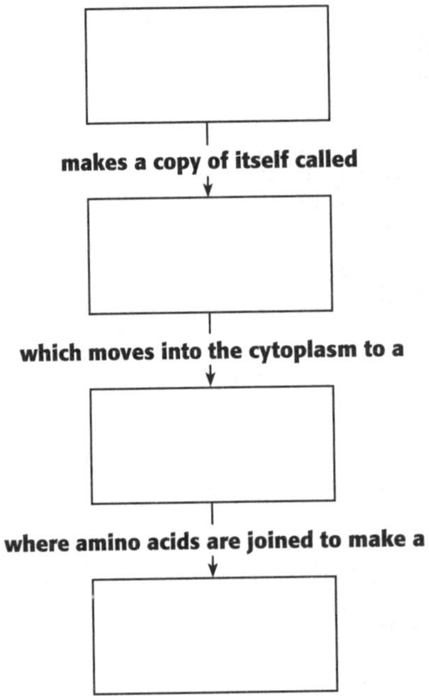
makes a copy of itself called
which moves into the cytoplasm to a
where amino acids are joined to make a

4. Draw Conclusions How can a mutation in a DNA base sequence cause a change in a gene and a trait? What determines whether the mutation is passed on to offspring?

____

____

____

5. Identify Give two ways that genetic fingerprinting can be used.

____

____

Name ______________________ Class ______________ Date ______________

SECTION 1

Change over Time

BEFORE YOU READ

After you read this section, you should be able to answer these questions:

- How are organisms different from one another?
- How do scientists know that species change over time?
- How can you tell if species are related?

National Science Education Standards
LS 2e, 3a, 3d, 4a, 5a, 5b, 5c

What Are Adaptations?

The pictures below show three different kinds of frogs. These frogs all have some things in common. For example, they all have long back legs to help them move. However, as you can see in the pictures, there are also many differences between the frogs. Each frog has some physical features that can help it survive.

The red-eyed tree frog has green skin. It hides in the leaves of trees during the day. It comes out at night.

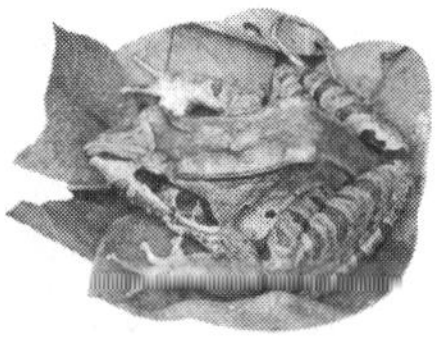

The skin of the smokey jungle frog looks like leaves. It can hide on the forest floor.

The strawberry poison frog has brightly colored skin. Its bright coloring warns predators that it is poisonous.

A feature that helps an organism survive and reproduce in its environment is called an **adaptation**. Some adaptations, such as striped skin or a long neck, are physical features. Other adaptations are behaviors that help an organism find food, protect itself, or reproduce.

Living things with the same features may be members of the same species. A **species** is a group of organisms that can mate with one another to produce fertile offspring. *Fertile* offspring are offspring that can reproduce. For example, all smokey jungle frogs are members of one species. Therefore, smokey jungle frogs can mate with each other to produce offspring. The offspring can also reproduce.

A group of individuals of the same species living in the same place is a *population*. For example, all of the smokey jungle frogs living in a certain jungle are a population.

STUDY TIP

Learn New Words As you read this section, circle the words you don't understand. When you figure out what they mean, write the words and their definitions in your notebook.

TAKE A LOOK

1. Explain How does the coloring of the strawberry poison frog help it survive?

____

____

____

____

Critical Thinking

2. Infer Give an example of a behavior that could help an organism survive.

____

____

____

CHANGE OVER TIME

Scientists estimate that there has been life on Earth for over 3 billion years. Earth has changed a great deal during its history. Living things have changed in this time too. Since life first appeared on Earth, many species have died out and many new species have appeared.

Scientists observe that species change over time. They also observe that the inherited characteristics in populations change over time. Scientists think that as populations change, new species form. New species descend from older species. The process in which populations change over time is called **evolution**. ☑

READING CHECK

3. Define What is evolution?

____

____

____

What Is the Evidence that Organisms Have Changed?

Much of the evidence that organisms have changed over time is buried in sedimentary rock. *Sedimentary rock* forms when pieces of sand, dust, or soil are laid down in flat layers. Sedimentary rocks may contain fossils. These fossils provide evidence that organisms have changed over Earth's history.

FOSSILS

Fossils are the remains or imprints of once-living organisms. Some fossils formed from whole organisms or from parts of organisms. Some fossils are signs, such as footprints, that an organism once existed. ☑

READING CHECK

4. Identify What are two kinds of fossils?

____

____

Fossils can form when layers of sediment cover a dead organism. Minerals in the sediment may seep into the organism and replace its body with stone. Fossils can also form when an organism dies and leaves an imprint of itself in sediment. Over time, the sediment can become rock and the imprint can be preserved as a fossil.

TAKE A LOOK

5. Compare Which of the fossils looks most like an organism that lives today? Give the modern organism that the fossil looks like.

____

____

Some fossils, like this trilobite, form from the bodies of organisms. The trilobite was an ancient marine animal.

Some fossils, like these ferns, form when an organism leaves an imprint in sediment. Over time, the sediment becomes rock and the imprint is preserved as a fossil.

THE FOSSIL RECORD

Comparing fossils provides evidence that organisms have changed over time. Rocks from different times in Earth's history contain fossils of different organisms. Fossils in newer layers of rock tend to be similar to present-day organisms. Fossils from older layers are less similar to present-day organisms. By studying fossils, scientists have made a timeline of life known as the **fossil record**. ☑

READING CHECK

6. Explain How do fossils give evidence that organisms have changed over time?

____

____

____

____

How Do Scientists Compare Organisms?

Scientists observe that all living things have some features in common. They also observe that all living things inherit features in a similar way. Therefore, scientists think that all living species are descended from a common ancestor. Scientists compare features of fossils and of living organisms to determine whether ancient species are related to modern species.

COMPARING STRUCTURES

When scientists study the anatomy, or structures, of different organisms, they find that some organisms share traits. These organisms may share a common ancestor. For example, the figure below shows that humans, cats, dolphins, and bats have similar structures in their front limbs. These similarities suggest that humans, cats, dolphins, and bats have a common ancestor.

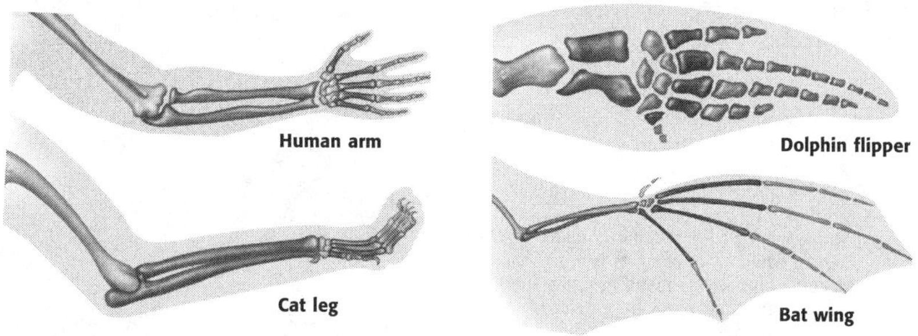
Human arm
Dolphin flipper
Cat leg
Bat wing

Comparing Structures The front limb bones of humans, cats, dolphins, and bats show some similarities. This suggests that all of these species share a common ancestor.

TAKE A LOOK

7. Identify Color the bones in each front limb to show which bones are similar. Use a different color for each bone.

Scientists can study the structures in modern organisms and compare them to structures in fossils. In this way, scientists can gather evidence that living organisms are related to organisms that lived long ago.

COMPARING CHEMICAL DATA

Remember that the genetic information in an organism's DNA determines the organism's traits. RNA and proteins also affect the traits of an organism. Scientists can compare these chemicals in organisms. The more alike these chemicals are between any two species, the more recently the two species shared a common ancestor.

Comparing DNA, RNA, and proteins can be very useful in determining whether species are related. However, this method can only be used on organisms that are alive today.

Critical Thinking

8. Infer Why can DNA, RNA, and proteins be used to compare only living organisms?

____

____

____

What Is the Evidence that Organisms Are Related?

Examining an organism carefully can give scientists clues about its ancestors. For example, whales look like some fish. However, unlike fish, whales breathe air, give birth to live young, and produce milk. These traits show that whales are mammals. Therefore, scientists think that whales evolved from ancient mammals. By examining fossils from ancient mammals, scientists have been able to determine how modern whales may have evolved. ☑

READING CHECK

9. Describe Why do scientists think that whales evolved from ancient mammals and not ancient fish?

____

____

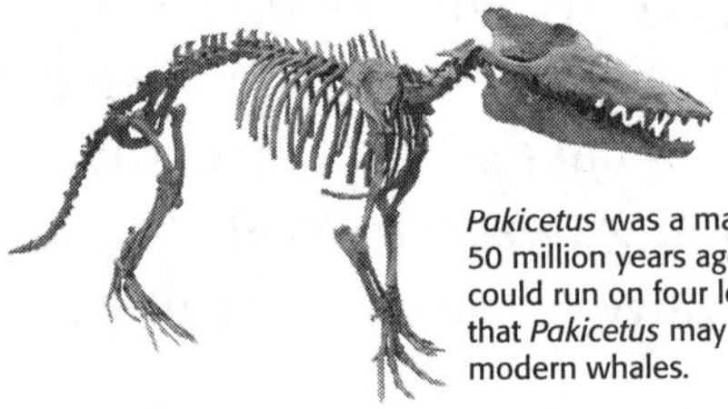

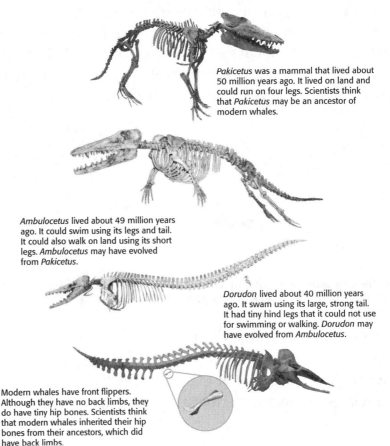

Pakicetus was a mammal that lived about 50 million years ago. It lived on land and could run on four legs. Scientists think that *Pakicetus* may be an ancestor of modern whales.

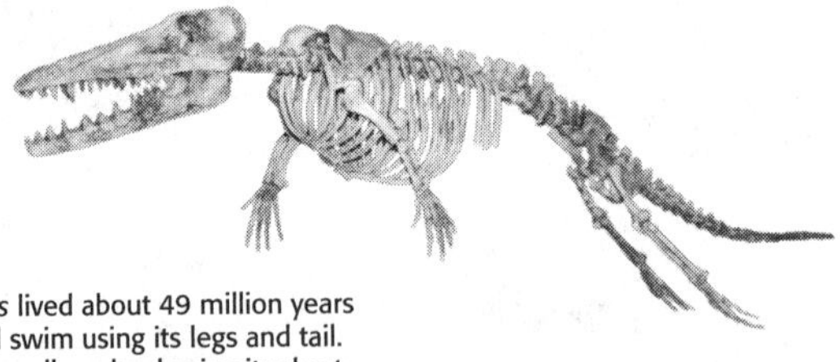

Ambulocetus lived about 49 million years ago. It could swim using its legs and tail. It could also walk on land using its short legs. *Ambulocetus* may have evolved from *Pakicetus*.

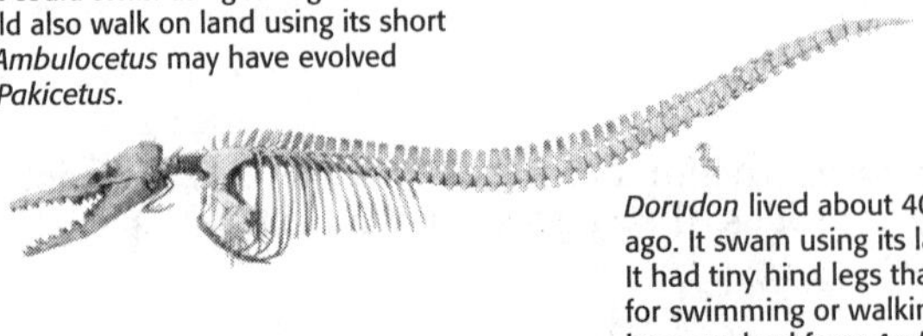

Dorudon lived about 40 million years ago. It swam using its large, strong tail. It had tiny hind legs that it could not use for swimming or walking. *Dorudon* may have evolved from *Ambulocetus*.

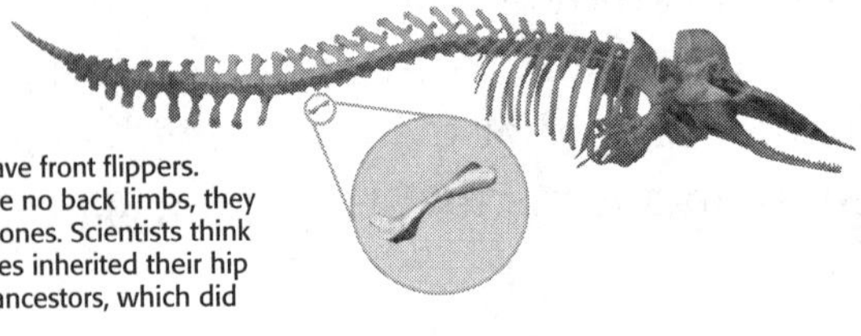

Modern whales have front flippers. Although they have no back limbs, they do have tiny hip bones. Scientists think that modern whales inherited their hip bones from their ancestors, which did have back limbs.

TAKE A LOOK

10. Compare Give two ways that modern whales are different from *Pakicetus*, and two ways that they are similar.

____

____

____

____

EVIDENCE FROM FOSSILS

Fossils provide several pieces of evidence that some species of ancient mammals are related to each other and to modern whales. First, each species shares some traits with an earlier species. Second, some species show new traits that are also found in later species. Third, each species had traits that allowed it to survive in a particular time and place in Earth's history.

EVIDENCE FROM MODERN WHALES

Some features of modern whales also suggest that they are related to ancient mammals. For example, modern whales do not have hind limbs. However, they do have tiny hip bones. Scientists think that modern whales inherited these hip bones from the whales' four-legged ancestors. Scientists often use this kind of evidence to determine the relationships between organisms.

How Do Scientists Show the Relationships Between Organisms?

As scientists analyze fossils and living organisms, they develop hypotheses about how species are related. They use *branching diagrams* to show the relationships between species.

This line represents an ancient species. This species is the common ancestor of the other species on the diagram.

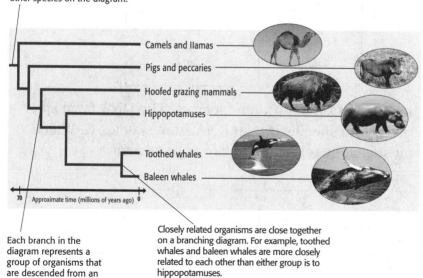

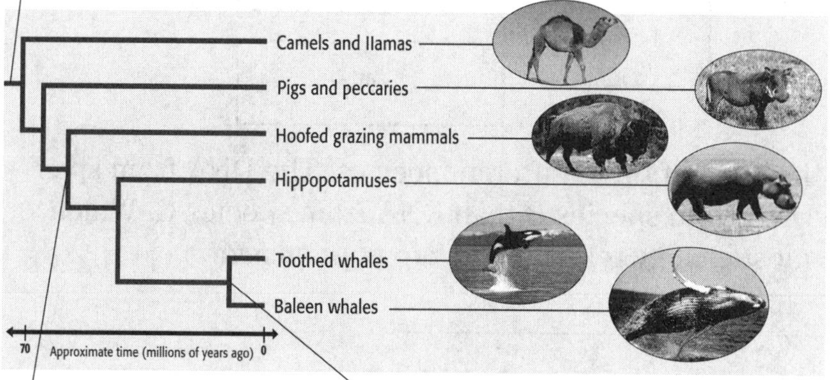
Camels and llamas
Pigs and peccaries
Hoofed grazing mammals
Hippopotamuses
Toothed whales
Baleen whales
70
Approximate time (millions of years ago)
0

Each branch in the diagram represents a group of organisms that are descended from an earlier species.

Closely related organisms are close together on a branching diagram. For example, toothed whales and baleen whales are more closely related to each other than either group is to hippopotamuses.

Scientists use branching diagrams to show how organisms are related.

STANDARDS CHECK

LS 5a Millions of species of animals, plants, and microorganisms are alive today. Although different species might look dissimilar, the unity among organisms becomes apparent from an analysis of internal structures, the similarity of their chemical processes, and the evidence of common ancestry.

Word Help: structure
the arrangement of the parts of a whole

Word Help: process
a set of steps, events, or changes

Word Help: evidence
information showing whether an idea or belief is true or valid

11. Identify Give one piece of evidence that indicates ancient mammals are related to modern whales.

____

____

TAKE A LOOK

12. Use a Model According to the diagram, which animals are most closely related to whales?

____

13. Use a Model According to the diagram, which animals are least closely related to whales?

____

Name ______________________ Class ______________ Date ______________

Section 1 Review

NSES LS 2e, 3a, 3d, 4a, 5a, 5b, 5c

SECTION VOCABULARY

adaptation a characteristic that improves an individual's ability to survive and reproduce in a particular environment

evolution the process in which inherited characteristics within a population change over generations such that new species sometimes arise

fossil the trace or remains of an organism that lived long ago, most commonly preserved in sedimentary rock

fossil record a historical sequence of life indicated by fossils found in layers of Earth's crust

species a group of organisms that are closely related and can mate to produce fertile offspring

1. Identify What evidence suggests that humans and bats have a common ancestor?

____

____

____

2. Make a Model Humans are most closely related to chimpanzees. Humans are least closely related to orangutans. Fill in the blank spaces on the branching diagram to show how humans, orangutans, chimpanzees, and gorillas are related.

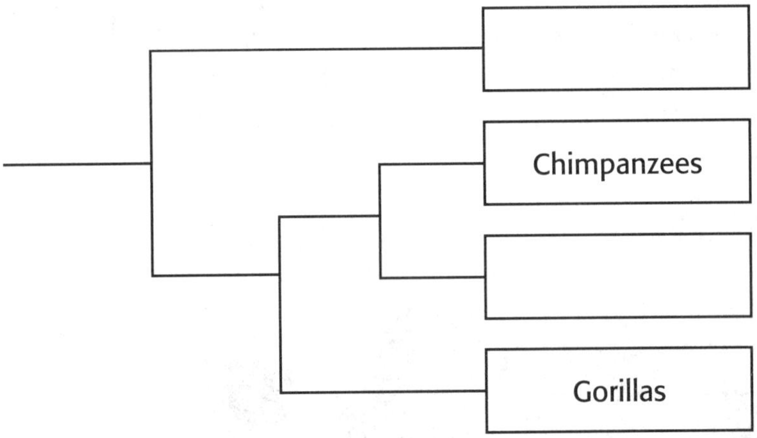
Chimpanzees
Gorillas

3. Infer A scientist studies the DNA of three different species. The DNA from species A is more similar to DNA from species B than DNA from species C. Which species are probably the most closely related? Explain your answer.

____

____

____

____

SECTION 2

How Does Evolution Happen?

BEFORE YOU READ

After you read this section, you should be able to answer these questions:

- Who was Charles Darwin?
- What ideas affected Darwin's thinking?
- What is natural selection?

National Science Education Standards
LS 2a, 2b, 2e, 3d, 5a, 5b

Who Was Charles Darwin?

In 1831, Charles Darwin graduated from college. Although he eventually earned a degree in religion, Darwin was most interested in the study of plants and animals.

Darwin's interest in nature led him to sign on for a five-year voyage around the world. He was a naturalist on the HMS *Beagle*, a British ship. A *naturalist* is someone who studies nature. During the trip, Darwin made observations that helped him form a theory about how evolution happens. These ideas caused scientists to change the way they thought about the living world.

STUDY TIP

Summarize After you read this section, make a chart showing the four steps of natural selection. In the chart, explain what happens at each step.

DARWIN'S JOURNEY

On the trip, Darwin observed plants and animals from many parts of the world. One place Darwin found interesting was the Galápagos Islands. These islands are located about 1,000 km west of Ecuador, a country in South America. Many unusual organisms live on the Galápagos Islands.

Math Focus

1. Convert About how far are the Galápagos Islands from Ecuador in miles? 1 km = 0.62 mi

____

This line shows the course of the HMS *Beagle*.

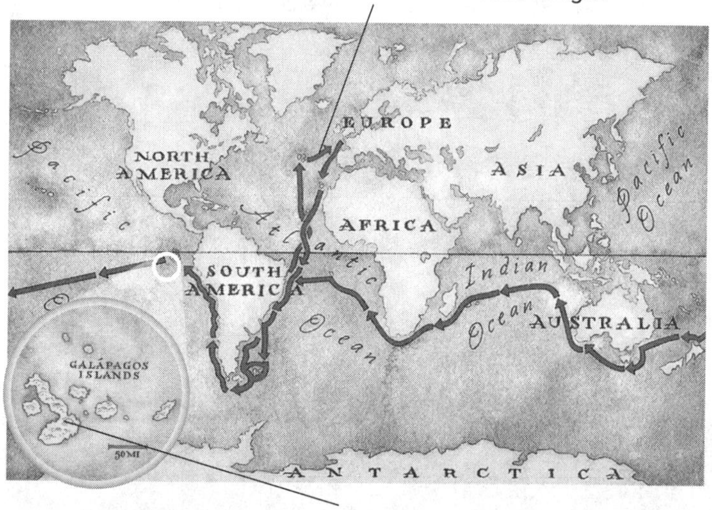
EUROPE
NORTH AMERICA
ASIA
Pacific Ocean
Atlantic Ocean
AFRICA
Pacific Ocean
SOUTH AMERICA
Indian Ocean
AUSTRALIA
GALÁPAGOS ISLANDS
50 MI
ANTARCTICA

Darwin studied plants and animals on the Galápagos Islands.

TAKE A LOOK

2. Describe Which continent are the Galápagos Islands closest to?

____

DARWIN'S FINCHES

Darwin observed that the animals and plants on the Galápagos Islands were similar to those in Ecuador. However, they were not identical. For example, Darwin closely observed birds called finches. The finches on the Galápagos Islands were slightly different from the finches in Ecuador. In addition, the finches on each island in the Galápagos differed from the finches on the other islands. ☑

READING CHECK

3. Describe What did Darwin observe about the finches on the Galápagos Islands?

Darwin hypothesized that the island finches were descendents of South American finches. He thought the first finches on the islands were blown there from South America by a storm. He suggested that over many generations, the finch populations evolved adaptations that helped them survive in the different island environments. For example, the beaks of different finch species are adapted to the kind of food the species eat.

The large ground finch has a wide, strong beak. It can easily crack open large, hard seeds. Its beak works like a nutcracker.

The cactus finch has a tough beak. It uses its beak to eat cactus parts and insects. Its beak works like a pair of needle-nose pliers.

The warbler finch has a small, narrow beak. It can catch small insects with its beak. Its beak works like a pair of tweezers.

Critical Thinking

4. Infer What can you guess about the environment in which the cactus finch lives based on the information in the figure? Explain your answer.

How Did Darwin Develop the Theory of Evolution by Natural Selection?

After Darwin returned to England, he spent many years thinking about his experiences on the trip. In 1859, Darwin published a famous book called *On the Origin of Species by Means of Natural Selection.* In his book, Darwin proposed the theory that evolution happens by natural selection.

Natural selection happens when organisms that are well adapted to their environment survive, but less well-adapted organisms do not. When the better-adapted organisms reproduce, they pass their useful traits on to their offspring. Over time, more members of the population have these traits. Darwin combined ideas about breeding, population, and Earth's history to come up with a theory to explain his observations. ☑

READING CHECK

5. Define What is natural selection?

____

____

____

____

____

IDEAS ABOUT BREEDING

In Darwin's time, farmers and breeders had produced many kinds of farm animals and plants. They learned that if they bred plants or animals that had a desirable trait, some of the offspring might have the trait. A **trait** is a form of an inherited characteristic. The practice in which humans select plants or animals for breeding based on desired traits is called **selective breeding**.

Selective breeding showed Darwin that the traits of organisms can change and that certain traits can spread through populations. For example, most pets, such as the dogs below, have been bred for a variety of desired traits. Over the past 12,000 years, people have selectively bred dogs to produce more than 150 breeds. ☑

READING CHECK

6. Explain How did ideas about selective breeding affect Darwin's thinking about evolution?

____

____

____

____

____

People have selectively bred dogs for different traits. Today, there are over 150 dog breeds.

IDEAS ABOUT POPULATION

During Darwin's time, a scientist named Thomas Malthus was studying human populations. He observed that there were more babies being born than there were people dying. He thought that the human population could grow more rapidly than food supplies could grow. This would result in a worldwide food shortage. Malthus also pointed out that the size of human populations is limited by problems such as starvation and disease. ☑

READING CHECK

7. Identify According to Thomas Malthus, what are two things that can limit the size of human populations?

____

____

Darwin realized that Malthus's ideas can apply to all species, not just humans. He knew that any species can produce many offspring. He also knew starvation, disease, competition, and predation limited the populations of all species. Only a limited number of individuals live long enough to reproduce.

Darwin reasoned that the survivors had traits that helped them survive in their environment. He also thought that the survivors would pass on some of their traits to their offspring.

IDEAS ABOUT EARTH'S HISTORY

New information about Earth's history also affected Darwin's ideas about evolution. During Darwin's time, most geologists thought that Earth was very young. But important books, such as *Principles of Geology* by Charles Lyell, were changing ideas about the Earth. Lyell's book gave evidence that Earth is much older than anyone once thought. ☑

READING CHECK

8. Explain How did Charles Lyell's book change how scientists thought about Earth's history?

____

____

____

Darwin thought that evolution happens slowly. Darwin reasoned that if Earth was very old, there would be enough time for organisms to change slowly.

Idea	How it contributed to Darwin's theory
Selective breeding	
	helped Darwin realize that not all of an organism's offspring will survive to reproduce
	helped Darwin realize that slow changes can produce large differences over a long period of time

TAKE A LOOK

9. Describe Fill in the blank spaces in the table.

HOW NATURAL SELECTION WORKS

Natural selection has four steps: *overproduction, inherited variation, struggle to survive,* and *successful reproduction.*

❶ Overproduction A tarantula's egg sac can hold 500 to 1,000 eggs. Some of the eggs will survive and develop into adult spiders. Some will not.

❷ Inherited Variation Every individual has its own combination of traits. Each tarantula is similar, but not identical, to its parents.

❸ Struggle to Survive Some tarantulas may have traits that make it more likely that they will survive. For example, a tarantula may be better able to fight off predators, such as this wasp.

❹ Successful Reproduction The tarantulas that are best adapted to their environment are likely to survive and reproduce. Their offspring may inherit the traits that help them to survive.

Say It

Give Examples The figure shows one example of how the four steps of natural selection can work. In a group, talk about three or more other examples of how natural selection can affect populations.

TAKE A LOOK

10. Identify Why are some tarantulas more likely to survive than others?

____

____

GENETICS AND EVOLUTION

Darwin knew that organisms inherit traits, but not how they inherit traits. He also knew that there is great variation among organisms, but not how that variation happens. Today, scientists know that genes determine the traits of an organism. These genes are exchanged and passed on from parent to offspring.

Name ______________________ Class ______________ Date ______________

Section 2 Review

NSES LS 2a, 2b, 2e, 3d, 5a, 5b

SECTION VOCABULARY

natural selection the process by which individuals that are better adapted to their environment survive and reproduce more successfully than less well adapted individuals do; a theory to explain the mechanism of evolution

selective breeding the human practice of breeding animals or plants that have certain desired traits

trait a genetically determined characteristic

1. **Explain** How did the ideas in Charles Lyell's book affect Darwin's thinking about evolution?

____

____

____

2. **Identify** In what way are the different finch species of the Galápagos Islands adapted to the different environments on the islands?

____

____

3. **Compare** How is natural selection different from selective breeding?

____

____

____

____

4. **Describe** How did Darwin apply Malthus's ideas about human populations to the theory of evolution by natural selection?

____

____

____

____

5. **List** What are the four steps of natural selection?

____

____

____

____

Name ______________________ Class ______________ Date ______________

CHAPTER 5 The Evolution of Living Things

SECTION 3

Natural Selection in Action

BEFORE YOU READ

After you read this section, you should be able to answer these questions:

- Why do populations change?
- How do new species form?

National Science Education Standards
LS 2a, 2e, 3d, 4d, 5b

Why Do Populations Change?

The theory of evolution by natural selection explains how changes in the environment can cause populations to change. Organisms that are well-adapted to their environment survive to reproduce. Organisms that are less well-adapted do not.

Some environmental changes are caused by people. Others happen naturally. No matter how they happen, though, environmental changes can cause populations to change.

STUDY TIP

Summarize As you read, underline the important ideas in each paragraph. When you finish reading, write a short summary of the section using the ideas that you underlined.

ADAPTATION TO HUNTING

Hunting is one of the factors that can affect the survival of animals. In Africa, people hunt male elephants for their tusks, which are made of ivory. Because of natural genetic variations, some male elephants do not grow tusks. People do not hunt these tuskless elephants, so tuskless elephants tend to live longer than elephants with tusks. Therefore, tuskless elephants are more likely to reproduce and pass the tuskless trait to their offspring. ☑

Over time, the tuskless trait has become more common. For example, in 1930, about 99% of male elephants in one area had tusks. Today, only about 85% of male elephants in that area have tusks.

READING CHECK

1. Explain Why are tuskless elephants more likely to reproduce in Africa?

____

____

People hunt elephants for the ivory in their tusks.

INSECTICIDE RESISTANCE

Many people use chemicals to control insect pests. These chemicals, called *insecticides*, kill insects. Sometimes, an insecticide that used to work well no longer affects an insect population. The insect population has evolved a resistance to the insecticide. This happens by natural selection, as shown in the figure below. ☑

READING CHECK

2. Define What is an insecticide?

____

____

❶ When it is first used, the insecticide kills most of the insects. However, a few insects have genes that make them resistant to the insecticide. These insects survive.

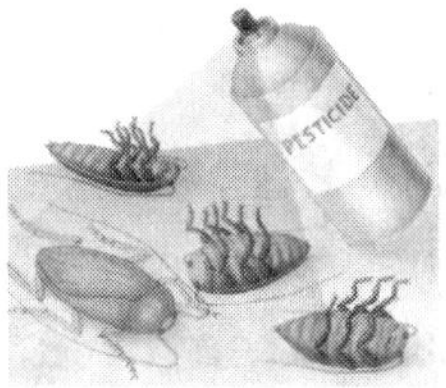

❷ The insects that are resistant to the insecticide pass on their genes to their offspring. Over time, almost all of the insects in the population have the insecticide-resistance gene.

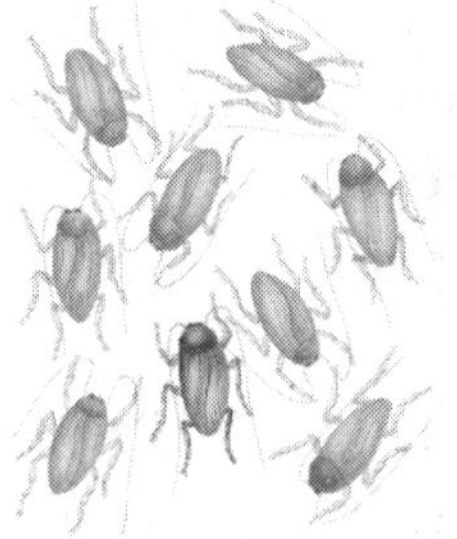

❸ When the same insecticide is used on the insects, only a few of the insects are killed. This is because most of the insects are resistant to the insecticide.

TAKE A LOOK

3. Infer What would happen to the population of insects if none were resistant to the insecticide?

____

Insect populations can evolve quickly. This happens for two reasons: insects have many offspring and they have a short generation time. **Generation time** is the average time between one generation and the next. In general, the longer the generation time for a population, the slower the population can evolve.

Critical Thinking

4. Apply Concepts The females of a certain species of mammal prefer to mate with less-colorful males. What will probably happen to the proportion of colorful males in the population with time?

____

____

COMPETITION FOR MATES

Organisms that reproduce sexually have to compete with one another for mates. For example, many female birds prefer to mate with colorful males. This means that colorful males have more offspring than less-colorful males. In most organisms, color is a genetic trait that is passed on to offspring. Therefore, colorful male birds are likely to produce colorful offspring. Over time, the proportion of colorful birds in the population will increase.

How Does Natural Selection Make New Species?

The formation of a new species as a result of evolution is called **speciation**. Three events often lead to speciation: separation, adaptation, and division. ☑

READING CHECK

5. Define Write the definition of speciation in your own words.

SEPARATION

Speciation may begin when a part of a population becomes separated from the rest. This can happen in many ways. For example, a newly formed canyon, mountain range, or lake can separate the members of a population.

ADAPTATION

After two groups have been separated, each group continues to be affected by natural selection. Different environmental factors may affect each population. Therefore, different traits can be favored in each population. Over many generations, different traits may spread through each population. ☑

READING CHECK

6. Explain Why may separated populations develop different traits?

DIVISION

Natural selection can cause two separated populations to become very different from each other. With time, the members of the two populations may be unable to mate successfully. The two populations may then be considered different species. The figure below shows how species of Galápagos finches may have evolved through separation, adaptation, and division.

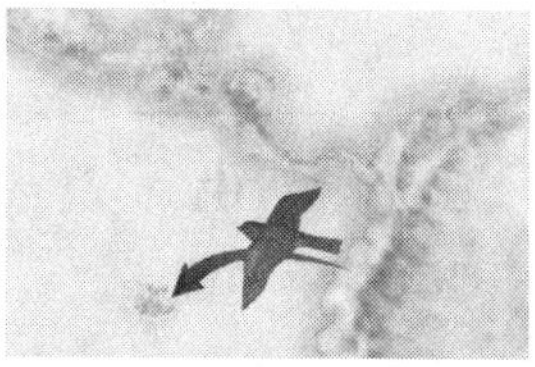

❶ **Separation** Some finches left the South American mainland and reached one of the Galápagos Islands.

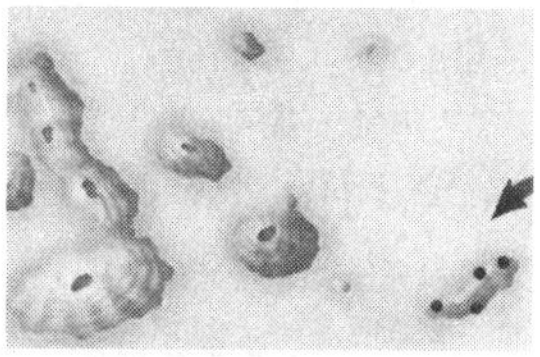

❷ **Adaptation** The finches on the island reproduced. Over time, they adapted to the environment on the island.

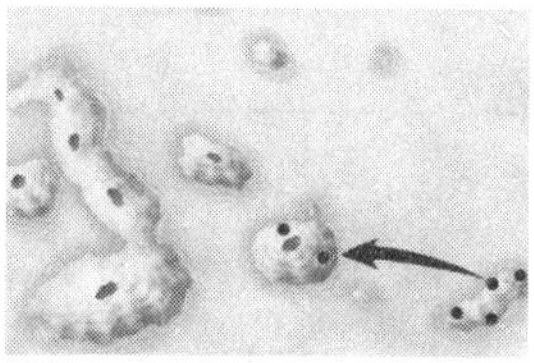

❸ **Separation** Some finches flew to a second island.

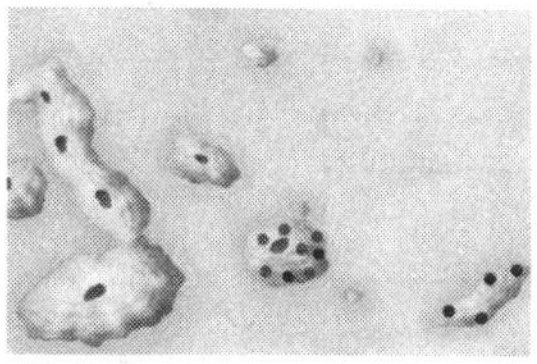

❹ **Adaptation** These finches reproduced on the second island. Over time, they adapted to the second island's environment.

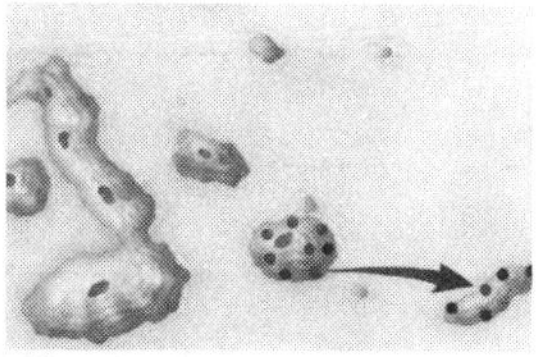

❺ **Division** After many generations, the finches on the second island were unable to successfully mate with the finches on the first island. The populations of finches on the two islands had become different species.

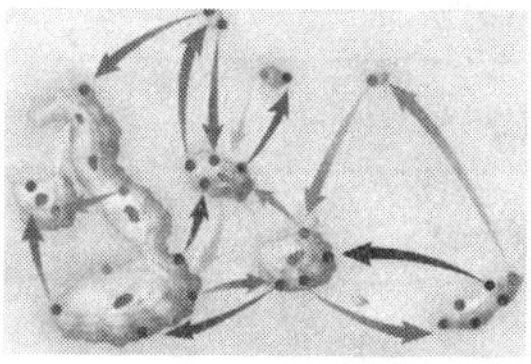

❻ **Speciation** This process may have happened many times as finches flew to the different islands in the Galápagos.

TAKE A LOOK

7. Identify Where did all of the finches on the Galápagos Islands originally come from?

Name ______________________ Class ______________ Date ______________

Section 3 Review

NSES LS 2a, 2e, 3d, 4d, 5b

SECTION VOCABULARY

generation time the period between the birth of one generation and the birth of the next generation	**speciation** the formation of new species as a result of evolution

1. **List** What are three events that can lead to speciation?

____

____

____

2. **Infer** What kinds of environmental factors may affect organisms that live on a rocky beach? Give three examples.

____

____

____

3. **Identify** Give three examples of things that can cause groups of individuals to become separated.

____

____

____

4. **Explain** Why can insects adapt to pesticides quickly? Give two reasons.

____

____

5. **Apply Concepts** Which of the organisms described below can probably evolve more quickly in response to environmental changes? Explain your answer.

Organism	Generation time	Average number of offspring per generation
A	6 years	50
B	2 years	100
C	10 years	5

____

____

____

Name ______________________ Class ______________ Date ______________

CHAPTER 6 The History of Life on Earth

SECTION 1

Evidence of the Past

BEFORE YOU READ

After you read this section, you should be able to answer these questions:

- What are fossils?
- What is the geologic time scale?
- How have conditions on Earth changed with time?

National Science Education Standards

LS 1a, 3d, 5b, 5c

What Are Fossils?

In 1995, geologist Paul Sereno found a dinosaur skull in a desert in Africa. The skull was 1.5 m long. The dinosaur it came from may have been the largest land predator that ever existed!

STUDY TIP

Organize After you read this section, make a Concept Map using the vocabulary terms from the section.

Scientists like Paul Sereno look for clues to help them learn what happened on Earth in the past. These scientists are called paleontologists. *Paleo* means "old," and *onto* means "life." Therefore, *paleontologists* study "old life," or life that existed in the past. One of the most important ways paleontologists learn about the past is by studying fossils.☑

READING CHECK

1. Define What is a paleontologist?

____

____

____

A **fossil** is the traces or remains of an organism that have been preserved over thousands of years. Some fossils form when an organism's body parts, such as shells or bones, are preserved in rock. Other fossils are signs, such as imprints, that an organism once existed. The figure below shows one way that this kind of fossil can form.

❶ An organism dies and is buried by sediment, such as mud or clay.

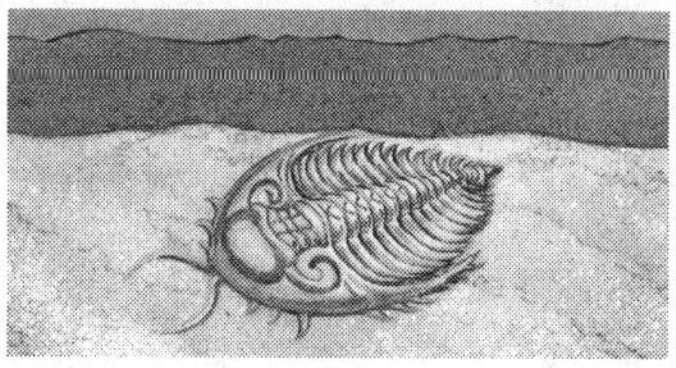

❷ After the organism dies, its body rots away. An imprint, or *mold*, is left in the sediment. The mold has the same shape as the organism's body.

❸ Over time, the mold fills with a different kind of sediment. The sediment forms a *cast* of the organism. The sediment can turn into rock, preserving the cast for millions of years.

TAKE A LOOK

2. Infer If a paleontologist found the cast of this fossil, would the paleontologist be able to tell what the inside of the organism looked like? Explain your answer.

____

____

____

____

How Do Scientists Know the Ages of Fossils?

There are two main methods that scientists use to learn how old a fossil is: relative dating and absolute dating. During **relative dating**, a scientist estimates how old a rock or fossil is compared to other rocks or fossils. ☑

READING CHECK

3. Define What is relative dating?

Almost all fossils are found in a kind of rock called *sedimentary rock*. Sedimentary rocks form in layers. In most bodies of sedimentary rock, the oldest layers are found at the bottom. The newest layers are found at the top. Scientists can use this information to estimate the relative ages of fossils in the layers. Fossils in the bottom layers are usually older than fossils in the top layers.

The second way that scientists learn the age of fossils is by absolute dating. During **absolute dating**, scientists learn the age of a fossil in years. Absolute dating is more precise than relative dating.

Critical Thinking

4. Compare How is absolute dating different from relative dating? Give two ways.

One kind of absolute dating uses atoms. *Atoms* are tiny particles that make up all matter. Some atoms are unstable and can *decay*, or break down. When an atom decays, it becomes a different kind of atom. It may also give off smaller particles, energy, or both. The new atom that forms is stable—it does not decay.

Each kind of unstable atom decays at a certain rate. This rate is called the atom's *half-life*. Scientists know the half-lives of many different kinds of unstable atoms. They can measure the ratio of these unstable atoms to stable atoms in a rock. They can use this ratio to calculate the age of the rock and any fossils in it.

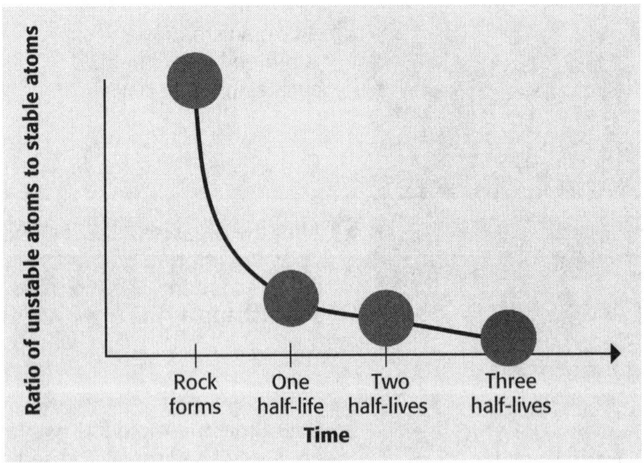
Ratio of unstable atoms to stable atoms
Rock forms
One half-life
Two half-lives
Three half-lives
Time

This figure shows how scientists can use half-lives to date rocks. After each half-life, the ratio of unstable to stable atoms in the rock decreases. Scientists can learn the age of the rock by measuring the ratio of unstable to stable atoms in it.

Math Focus

5. Read a Graph How does the ratio of unstable to stable atoms change over time?

What Is the Geologic Time Scale?

Think about some important things that have happened to you. You probably know the day, month, and year in which many of these events happened. The divisions of the calendar make it easy for you to remember these things. They also make it easy for you to talk to someone else about things that happened in the past.

In a similar way, geologists use a type of calendar to divide the Earth's history. This calendar is called the **geologic time scale**. The Earth is very old. Therefore, the divisions of time in the geologic time scale are very long.

Geologists use the geologic time scale to help make sense of how life on Earth has changed. When a new fossil is discovered, geologists figure out how old it is. Then, they place it in the geologic time scale at the correct time. In this way, they can construct a history of life on Earth. The figure below shows part of the geologic time scale.

Era	Period	Millions of years ago
Cenozoic era 	Quaternary	1.8
	Tertiary	65.5
Mesozoic era	Cretaceous 	146
	Jurassic 	200
	Triassic	251
Paleozoic era 	Permian 	299
	Carboniferous	359
	Devonian	416
	Silurian	444
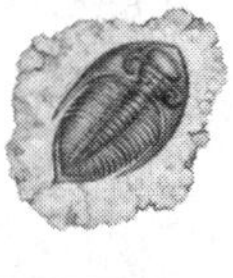	Ordovician	488
	Cambrian	542
Precambrian time 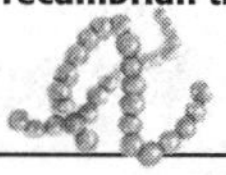 		4,600

Critical Thinking

6. Predict Consequences All geologists use the same basic geologic time scale. What might happen if every geologist used a different geologic time scale to study the Earth's history? Explain your answer.

____
____
____
____
____
____
____

Say It

Investigate Many geologic time-scale divisions are named after certain places on Earth. Research one of the periods in the Paleozoic era to find out where its name comes from. Share your findings with a small group.

TAKE A LOOK

7. List Give the three periods in the Mesozoic era in order from oldest to most recent.

____
____
____

STANDARDS CHECK

LS 5c Extinction of a species occurs when the environment changes and the adaptive characteristics of a species are insufficient to allow its survival. Fossils indicate that many organisms that lived long ago are extinct. Extinction of species is common; most of the species that have lived on Earth no longer exist.

Word Help: occur
to happen

Word Help: environment
the surrounding natural conditions that affect an organism

Word Help: insufficient
not enough

Word Help: survival
the continuing to live or exist; the act of continuing to live

Word Help: indicate
to be or give a sign of; to show

8. Infer How could global cooling cause a species to go extinct?

____

____

____

____

How Do Geologists Divide the Geologic Time Scale?

Many of the divisions of the geologic time scale are based on information from fossils. As scientists discover new fossils, they may add information to the geologic time scale. For example, most fossils that have been found are from organisms that have lived since Precambrian time. Therefore, little is known about life on Earth before this time. As scientists find more fossils from Precambrian time, more information may be added to the time scale.

Most of the divisions in the geologic time scale mark times when life on Earth changed. During many of these times, species died out completely, or became **extinct**. A *mass extinction* happens when many species go extinct at the same time.

Scientists do not know what causes all mass extinctions. The table below gives some events that may cause mass extinctions.

Event	How it causes mass extinction
Comet or meteorite impact	can throw dust and ash into the atmosphere, causing global cooling
Many volcanic eruptions	can throw dust and ash into the atmosphere, causing global cooling
Changing ocean currents	can cause climate change by changing how heat is distributed on Earth's surface

How Has the Earth's Surface Changed?

Geologists have found fossils of tropical plants in Antarctica. These kinds of plants cannot live in Antarctica today because the climate is too cold. The fossils show that Antarctica's climate must once have been warmer.

In the early 1900s, German scientist Alfred Wegener studied the shapes of the continents and the fossils found on them. From his observations, he proposed a hypothesis that all the continents were once joined together. They formed a large land mass called *Pangaea*. Since then, the continents have moved over the Earth's surface to their present locations.

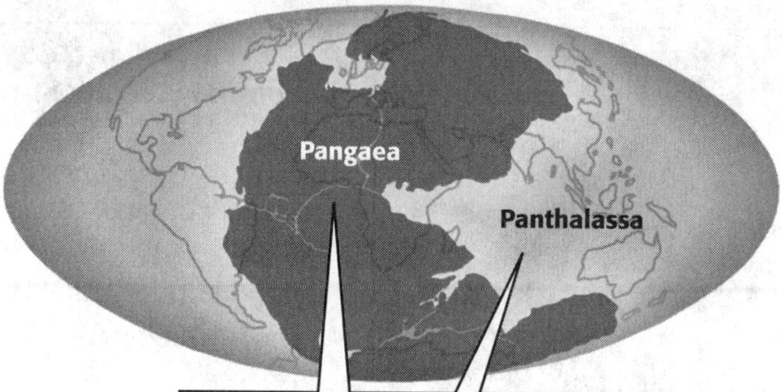
Pangaea
Panthalassa

About 245 million years ago, all of the continents were joined into one large land mass called *Pangaea*. It was surrounded by a huge sea called *Panthalassa*, which is Greek for "all seas."

TAKE A LOOK

9. Infer On the map, circle the part of Pangaea that probably became the continent of Africa.

MOVING CONTINENTS

In the 1960s and 1970s, new technology allowed geologists to learn more about the Earth's crust. They found that the crust is not one solid chunk of rock. Instead, it is broken into many pieces called *tectonic plates*. These plates move slowly over the Earth's surface. Some of the plates have continents on them. As the plates move, the continents are carried along. The theory of how the plates move is called the theory of **plate tectonics**. ☑

READING CHECK

10. Explain Why do continents move?

____

____

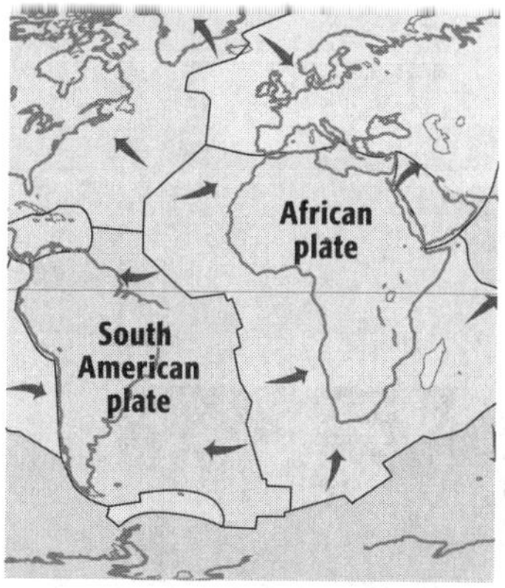
African plate
South American plate

The continents ride on tectonic plates. The plates move slowly over the Earth's surface.

The theory of plate tectonics explains why fossils of tropical plants have been found in Antarctica. Millions of years ago, Antarctica was located near the equator. Over time, the plate containing Antarctica has moved to its current location.

EFFECTS OF PLATE MOTIONS ON LIVING THINGS

Sudden changes on Earth's surface, such as a meteorite impact, may cause mass extinctions. Slower changes, such as tectonic plate movements, give populations of organisms time to adapt. Geologists use fossils of organisms to study how populations have adapted to changes on Earth.

Name ______________________ Class ______________ Date ______________

Section 1 Review

NSES LS 1a, 3d, 5b, 5c

SECTION VOCABULARY

absolute dating any method of measuring the age of an event or object in years

extinct describes a species that has died out completely

fossil the trace or remains of an organism that lived long ago, most commonly preserved in sedimentary rock

geologic time scale the standard method used to divide the Earth's long natural history into manageable parts

plate tectonics the theory that explains how large pieces of the Earth's outermost layer, called tectonic plates, move and change shape

relative dating any method of determining whether an event or object is older or younger than other events or objects

1. **Define** Write your own definition for *geologic time scale.*

____

____

2. **List** Give three things that can cause mass extinctions.

____

____

____

3. **Compare** Which fossil in the rock layers below is probably the oldest? Explain your answer.

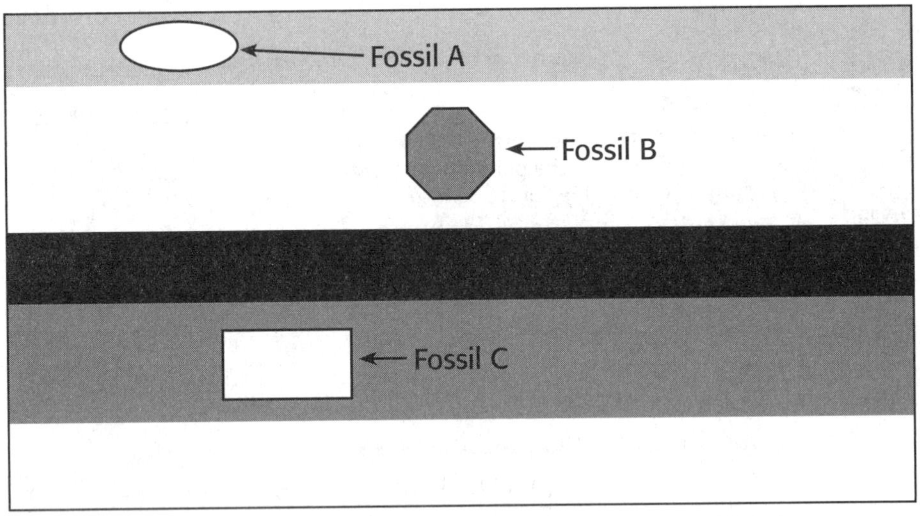

Fossil A
Fossil B
Fossil C

____

4. **Explain** Describe one way that a fossil can form.

____

____

____

Name ______________________ Class ______________ Date ______________

CHAPTER 6 The History of Life on Earth

SECTION 2 Eras of the Geologic Time Scale

BEFORE YOU READ

After you read this section, you should be able to answer these questions:

- What kinds of organisms evolved during Precambrian time?
- What kinds of organisms evolved during the Paleozoic, Mesozoic, and Cenozoic eras?

National Science Education Standards

LS 1a, 1b, 3d, 5a, 5b, 5c

What Happened During Precambrian Time?

Remember that geologists divide the Earth's history into four main parts. These parts are Precambrian time, the Paleozoic era, the Mesozoic era, and the Cenozoic era. During each part of Earth's history, different species arose and evolved.

Precambrian time was the longest part of Earth's history. It lasted from the time the Earth formed, 4.6 billion years ago, until about 542 million years ago. Life on Earth began during this time.

Scientists think that the early Earth was very different than the Earth today. On the early Earth, the atmosphere contained very little oxygen. It was made mostly of carbon dioxide, water vapor, and nitrogen gas. In addition, volcanic eruptions, meteorite and comet impacts, and severe storms were common. There was no ozone layer, so more ultraviolet radiation reached the Earth's surface.

Scientists think that life developed from simple chemicals in the oceans and the atmosphere. According to this hypothesis, energy from radiation and storms caused these chemicals to react. The reactions produced more complex molecules. These complex molecules combined to produce structures such as cells. ☑

This is what the early Earth may have looked like. An artist drew this picture based on information from scientists.

STUDY TIP

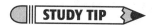

Organize As you read this section, make a chart showing the kinds of organisms that evolved during each of the four main parts of Earth's history.

Math Focus

1. Calculate About how long did Precambrian time last? Round your answer to the nearest hundred million years.

____

READING CHECK

2. Identify How do scientists think complex molecules formed on the early Earth?

____

____

____

PHOTOSYNTHESIS AND OXYGEN

The first living things on Earth were probably *prokaryotes*, or single-celled organisms without nuclei. These early organisms did not need oxygen to survive. However, more than 3 billion years ago, organisms called *cyanobacteria* began to develop. Cyanobacteria perform photosynthesis. During *photosynthesis*, a living thing uses sunlight, water, and carbon dioxide to produce food and oxygen. ☑

As the cyanobacteria carried out photosynthesis, they released oxygen into the atmosphere. This caused the amount of oxygen in the atmosphere to increase.

Some of the oxygen formed the ozone layer in the upper atmosphere. The ozone layer absorbs much of the ultraviolet radiation from the sun. Ultraviolet radiation is harmful to most living things. Before the ozone layer formed, life existed only in the oceans and underground. The new ozone layer reduced the radiation reaching Earth's surface. As a result, organisms were able to live on land.

READING CHECK

3. Define What is photosynthesis?

____

____

____

____

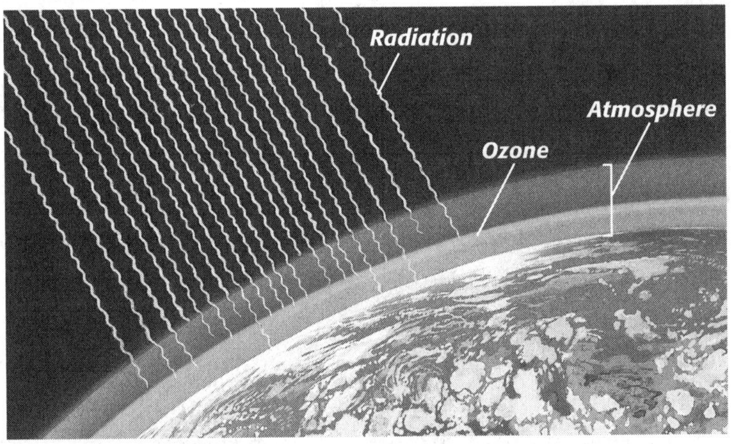
Radiation
Atmosphere
Ozone

Cyanobacteria released oxygen into the atmosphere during photosynthesis. Some of this oxygen formed the ozone layer, which absorbs much of the harmful radiation from the sun.

TAKE A LOOK

4. Explain Where did the oxygen come from that formed the ozone layer?

____

____

____

MULTICELLULAR ORGANISMS

About 2 billion years ago, single-celled organisms that were larger and more complex than prokaryotes evolved. These organisms were *eukaryotes*. They had nuclei and other complex structures in their cells. These early eukaryotic cells probably evolved into organisms with many cells.

What Happened During the Paleozoic Era?

The **Paleozoic era** began about 542 million years ago. It ended about 251 million years ago. *Paleozoic* comes from Greek words meaning "ancient life." The Paleozoic era is the second-longest part of the Earth's history. However, it was less than 10% as long as Precambrian time.

Era	Period	Millions of years ago
Paleozoic era 	Permian 	299
	Carboniferous	359
	Devonian	416
	Silurian	444
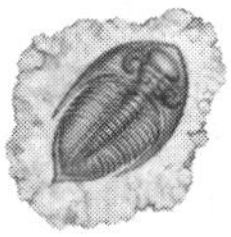	Ordovician	488
	Cambrian	542

TAKE A LOOK

5. List Give the six periods in the Paleozoic era in order from oldest to most recent.

______
______
______
______
______
______

LIFE IN THE PALEOZOIC ERA

Rocks from the Paleozoic era contain many fossils of *multicellular* organisms, or organisms with many cells. These organisms all evolved during the Paleozoic era. Most of the organisms that lived during the early Paleozoic era lived in the oceans. Some of the animals include sponges, corals, snails, clams, fishes, and sharks. ☑

During the Paleozoic era, plants and animals began to live on land. By the end of the era, forests of giant ferns, club mosses, and conifers covered much of the Earth. All of the major plant groups except for flowering plants developed during the Paleozoic. These plants provided food and shelter for animals.

Fossils indicate that crawling insects were some of the first animals to live on land. By the end of the Paleozoic, amphibians, reptiles, and winged insects had evolved. ☑

READING CHECK

6. Identify Where did most animals live during the early Paleozoic?

______

READING CHECK

7. Identify What kinds of animals were probably the first to live on land?

______

THE END OF THE PALEOZOIC

The end of the Paleozoic is marked by the largest known mass extinction. During this extinction, as many as 90% of marine species died out. Scientists are not sure what caused this extinction.

Era	Period	Millions of years ago
Mesozoic era	 Cretaceous	146
	 Jurassic	200
	Triassic	251

What Happened During the Mesozoic Era?

The **Mesozoic era** lasted from about 251 million years ago to about 65.5 million years ago. *Mesozoic* comes from Greek words that mean "middle life." Many species of reptiles evolved during the Mesozoic era. They became the dominant, or main, organisms on Earth. The Mesozoic era is sometimes called the "Age of Reptiles." ☑

READING CHECK

8. Explain Why is the Mesozoic era sometimes called the "Age of Reptiles"?

____

____

____

LIFE IN THE MESOZOIC ERA

Dinosaurs are the most well-known reptiles that evolved during the Mesozoic. They dominated the Earth for about 150 million years. As you may know, there were many different species of dinosaurs. Some were taller than a 10-story building. Others were as small as modern chickens.

In addition to dinosaurs, the first birds and mammals evolved during the Mesozoic. In fact, many scientists think that birds evolved from some kinds of dinosaurs.

The most important land plants during the early Mesozoic era were conifers. These conifers formed huge forests. Flowering plants evolved later in the Mesozoic.

Critical Thinking

9. Infer Scientists sometimes call birds "living dinosaurs." Why do you think this is?

____

____

____

THE END OF THE MESOZOIC ERA

About 65.5 million years ago, dinosaurs and many other species became extinct. Most scientists think that this extinction was caused by the impact of a large meteorite or comet. ☑

The impact produced huge amounts of dust and ash. This blocked sunlight from reaching Earth's surface. It also caused global temperatures to drop for many years. Without sunlight, plants began to die. Plant-eating dinosaurs soon died out because they had nothing to eat. Then, meat-eating dinosaurs died out.

READING CHECK

10. Identify What do most scientists think caused the extinction of the dinosaurs?

____

What Has Happened During the Cenozoic Era?

The **Cenozoic era** began about 65.5 million years ago and continues today. *Cenozoic* comes from Greek words meaning "recent life." During the Cenozoic era, mammals have become the dominant organisms. Therefore, the Cenozoic era is sometimes called the "Age of Mammals."

Era	Period	Millions of years ago
Cenozoic era 	Quaternary	1.8
	Tertiary	65.5

TAKE A LOOK

11. Identify How long ago did the Cenozoic era start?

____

Scientists have more information about the Cenozoic era than about any of the other parts of Earth's history. Fossils of organisms from this era formed more recently than fossils from earlier times. Therefore, most fossils of organisms from the Cenozoic era are near the surface. This makes them easier to find. In addition, the fossils from the Cenozoic era are less likely to have been destroyed by weathering and erosion. ☑

READING CHECK

12. Describe Give two reasons that scientists know more about the Cenozoic era than about other times in Earth's history.

____
____
____
____
____
____

LIFE IN THE CENOZOIC ERA

During the early parts of the Cenozoic era, most mammals were small. They lived in forests. Later in the era, larger mammals evolved. These included mastodons, saber-toothed cats, camels, and horses. Some had long legs for running. Some had special teeth for eating certain kinds of food. Some had large brains.

We are currently living in the Cenozoic era. Humans evolved during this era. The environment, landscapes, and organisms around us today are part of this era.

The climate has changed many times during the Cenozoic era. During some periods of time, the climate was much colder than it is today. These periods of time are known as *ice ages*. During ice ages, ice sheets and glaciers grew larger. Many organisms migrated toward the equator to survive the cold. Others adapted to the cold. Some species became extinct.

Name ______________________ Class ______________ Date ______________

Section 2 Review

NSES LS 1a, 1b, 3d, 5a, 5b, 5c

SECTION VOCABULARY

Cenozoic era the current geologic era, which began 65.5 million years ago; also called the Age of Mammals

Mesozoic era the geologic era that lasted from 251 million to 65.5 million years ago; also called the Age of Reptiles

Paleozoic era the geologic era that followed Precambrian time and that lasted from 542 million to 251 million years ago

Precambrian time the interval of time in the geologic time scale from Earth's formation to the beginning of the Paleozoic era, from 4.6 billion to 542 million years ago

1. **List** Give the four main parts of Earth's history in order from oldest to most recent.

____

____

2. **Identify** During which era did multicellular organisms evolve?

____

3. **Compare** Give three ways the early Earth was different from the Earth today.

____

____

____

4. **Describe** How did the Earth's atmosphere change because of cyanobacteria?

____

____

____

5. **Identify** What kind of event marks the end of both the Paleozoic era and the Mesozoic era?

____

6. **Infer** Why might birds and mammals have been more likely than reptiles to survive the events that caused the extinction of the dinosaurs?

____

____

7. **Describe** What happens during an ice age?

____

____

Name ________________________ Class ______________ Date ____________

CHAPTER 6 The History of Life on Earth

SECTION 3

Humans and Other Primates

BEFORE YOU READ

After you read this section, you should be able to answer these questions:

- What are two features of primates?
- How have hominids changed through time?

National Science Education Standards

LS 1a, 3d, 5a, 5b, 5c

How Are Humans Similar to Apes?

Humans, apes, and monkeys share many common features. In addition, scientists have found many fossils of organisms with features of both humans and apes. Therefore, scientists agree that humans, apes, and monkeys share a common ancestor. The species of humans, apes, and monkeys that are alive today all evolved from a single, earlier species. This species probably lived more than 45 million years ago.

Humans, monkeys, apes, and lemurs are all part of a group of mammals called **primates**. There are two features that primates share. First, a primate's eyes are located at the front of its head. Both eyes look in the same direction. This gives the primate *binocular*, or three-dimensional, vision.

A second feature primates share is flexible fingers. Almost all primates have *opposable thumbs*. This means that the thumb can move to touch each finger. It is not fixed in place like the toes of a dog or cat. Opposable thumbs help primates grip objects firmly.

◀ A primate's eyes both point in the same direction. This gives the primate three-dimensional vision.

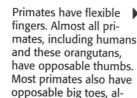

Primates have flexible fingers. Almost all primates, including humans and these orangutans, have opposable thumbs. Most primates also have opposable big toes, although humans do not. ▶

STUDY TIP

Summarize As you read this section, make a timeline that shows how hominids have changed over time.

STANDARDS CHECK

LS 5a Millions of species of animals, plants, and microorganisms are alive today. Although different species might look dissimilar, the unity among organisms becomes apparent from an analysis of internal structures, the similarity of their chemical processes, and the evidence of common ancestry.

Word Help: structure
a whole that is built or put together from parts

Word Help: process
a set of steps, events, or changes

Word Help: evidence
information showing whether an idea or belief is true or valid

1. Identify What are two features that humans share with apes and monkeys?

____

____

____

THE FIRST PRIMATES

The ancestors of primates probably lived at the same time as the dinosaurs. They probably lived in trees and ate insects. They were small and looked similar to mice. The first primates evolved during the early Cenozoic era. About 45 million years ago, primates with larger brains evolved. These primates were the first to share features with monkeys, apes, and humans. ☑

READING CHECK

2. Identify When did the first primates evolve?

____

____

APES AND CHIMPANZEES

The chimpanzee is a kind of ape. Scientists think that the chimpanzee is the closest living relative of humans. This does not mean that humans evolved from chimpanzees. It means that humans and chimpanzees share a more recent common ancestor than humans and other apes. Between 30 million and 6 million years ago, ancestors of humans, chimpanzees, and other apes began to evolve different features. ☑

READING CHECK

3. Identify What organism do scientists think is the closest living relative of humans?

____

HOMINIDS

Humans are part of a group of primates called hominids. The **hominid** family includes only humans and their human-like ancestors. Humans are the only species of hominids that are still living.

The main feature that separates hominids from other primates is bipedalism. *Bipedalism* means "walking on two feet." Humans are bipedal. Other primates are not bipedal. They mainly move around with all four limbs touching the ground.

The skeletons of humans and gorillas are similar. However, they are not exactly the same.

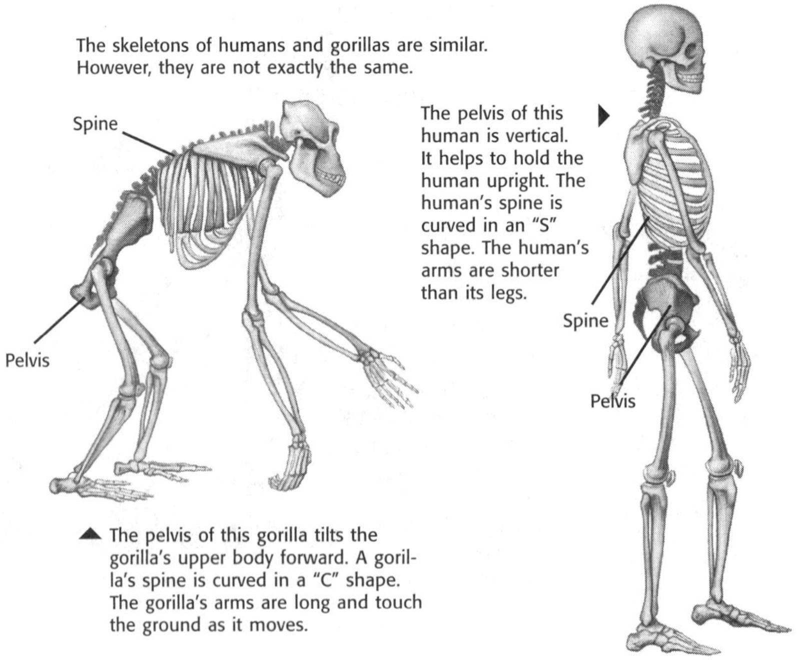
Spine
Pelvis
Spine
Pelvis

▲ The pelvis of this gorilla tilts the gorilla's upper body forward. A gorilla's spine is curved in a "C" shape. The gorilla's arms are long and touch the ground as it moves.

The pelvis of this human is vertical. It helps to hold the human upright. The human's spine is curved in an "S" shape. The human's arms are shorter than its legs. ▶

TAKE A LOOK

4. Compare How is the shape of a human spine different from that of a gorilla?

____

____

____

How Have Hominids Changed Through Time?

Scientists continue to find fossils of hominids. These fossils help scientists piece together the hominid family tree. As scientists find more fossils, they will better understand how modern humans evolved.

THE EARLIEST HOMINIDS

The earliest hominids had some humanlike features. Like all hominids, they could walk upright. They also had smaller teeth, flatter faces, and larger brains than earlier primates.

The oldest hominid fossils are more than 7 million years old. They have been found in Africa. Therefore, scientists think that hominids first evolved in Africa. ☑

READING CHECK

5. Explain Why do scientists think that hominids first evolved in Africa?

____

____

These are footprints from an early hominid. They are about 3.6 million years old. Anthropologist Mary Leaky discovered these footprints in Tanzania, Africa.

TAKE A LOOK

6. Apply Concepts Did the hominid that made these prints walk on two feet or four? Explain your answer.

____

____

Scientists have classified many early hominids as australopithecines. *Australopithecines* were apelike hominids. They had slightly larger brains than apes and may have used simple stone tools.

A VARIETY OF EARLY HOMINIDS

Many australopithecines and other types of hominids lived at the same time, just as many species of apes live today. Some australopithecines had slender bodies. They had humanlike jaws and teeth but small, apelike skulls. They probably lived in forests and grasslands. Scientists think that some of these australopithecines were the ancestors of modern humans. ☑

READING CHECK

7. Identify What type of hominid do scientists think was the ancestor of modern humans?

____

Some early hominids had large bodies and massive teeth and jaws. They had fairly small brains. They probably lived in tropical forests. Scientists think that these large-bodied hominids were probably not the ancestors of modern humans.

THE GROUP *HOMO*

About 2.4 million years ago, a new group of hominids evolved. These hominids were similar to the slender australopithecines, but were more humanlike. They had larger and more complex brains, rounder skulls, and flatter faces. They were probably scavengers, eating many different kinds of food. They may have migrated or changed the way they lived to adapt to climate change. ☑

READING CHECK

8. Compare How were the hominids that evolved 2.4 million years ago different from earlier australopithecines?

____

____

____

These new hominids were members of the group *Homo*, which includes modern humans. Fossil evidence shows that several different species of *Homo* may have lived at the same time on several continents. One of these species was *Homo habilis*, which evolved about 2.4 million years ago. Another species, *Homo erectus*, evolved about 1.8 million years ago. Members of the *Homo erectus* species could grow as tall as modern humans.

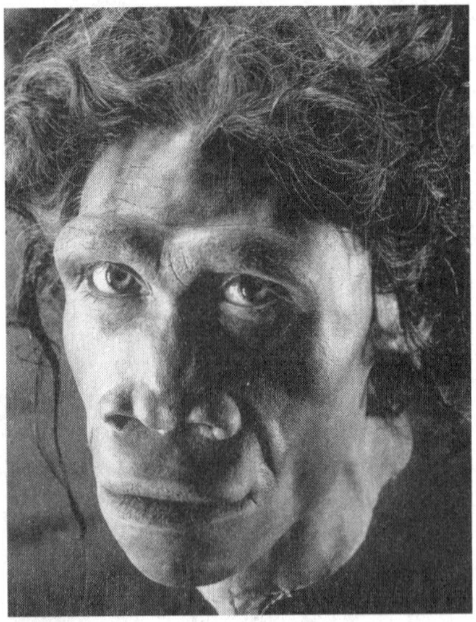

This is an artist's idea of what *Homo erectus* looked like. Artists and scientists work together to produce models like this. They used information from bones to infer what *Homo erectus* may have looked like.

TAKE A LOOK

9. Explain How do scientists infer what early hominids looked like?

____

____

When Did Modern Humans Evolve?

Until about 30,000 years ago, two types of hominids may have lived in the same areas at the same time. Both had large brains and made advanced tools, clothing, and art. Scientists think that one of these early hominids was the same species as modern humans.

Critical Thinking

10. Infer What kinds of evidence may scientists have used to determine that early hominids used tools and wore clothing?

____

____

____

____

NEANDERTHALS

One type of recent hominid is known as the *Neanderthal.* Neanderthals lived in Europe and western Asia. They may have evolved as early as 230,000 years ago. Neanderthals hunted large animals, made fires, and wore clothing. They may have cared for their sick and buried their dead. About 30,000 years ago, Neanderthals disappeared. Scientists do not know why they went extinct.

HOMO SAPIENS

Modern humans are members of the species **Homo sapiens**. The earliest members of *Homo sapiens* probably evolved in Africa 150,000 to 100,000 years ago. Some members of this species migrated out of Africa between 100,000 and 40,000 years ago. Compared to Neanderthals, *Homo sapiens* have smaller and flatter faces and more rounded skulls. *Homo sapiens* is the only species of hominid that is still alive. ☑

READING CHECK

11. Identify How many species of *Homo* are still alive?

____

Like modern humans, early *Homo sapiens* produced large amounts of art. They made sculptures, carvings, and paintings. Scientists have also found preserved villages and burial grounds from early *Homo sapiens.* These remains show that these early humans had an organized and complex society, like humans today.

This photo shows a museum recreation of early *Homo sapiens.*

Name ______________________ Class ______________ Date ______________

Section 3 Review

NSES **LS 1a, 3d, 5a, 5b, 5c**

SECTION VOCABULARY

hominid a type of primate characterized by bipedalism, relatively long lower limbs, and lack of a tail; examples include humans and their ancestors	***Homo sapiens*** the species of hominids that includes modern humans and their closest ancestors and that first appeared about 100,000 to 150,000 years ago **primate** a type of mammal characterized by opposable thumbs and binocular vision

1. Identify What feature separates hominids from other primates?

____

2. Identify When did humans, chimpanzees, and other apes begin to evolve different features?

____

3. Describe Give three features of the australopithecines that scientists think evolved into modern humans.

____

____

____

4. List Give three species of *Homo* and tell when each evolved.

____

____

____

5. Compare How are *Homo sapiens* different from Neanderthals? Give two ways.

____

____

6. Explain How do scientists know that many species of hominids once existed?

____

7. Infer What advantages could larger brains have given hominids? Give three examples.

____

____

____

Name ______________________ Class ______________ Date ______________

CHAPTER 7 Classification

SECTION 1

Sorting It All Out

BEFORE YOU READ

After you read this section, you should be able to answer these questions:

- What is classification?
- How do scientists classify organisms?
- How do scientists name groups of organisms?

National Science Education Standards
LS 5a

Why Do We Classify Things?

Imagine that you lived in a tropical rain forest and had to get your own food, shelter, and clothing from the forest. What would you need to know to survive? You would need to know which plants were safe to eat and which were not. You would need to know which animals you could eat and which ones could eat you. In other words, you would need to study the organisms around you and put them into useful groups. You would *classify* them.

Biologists use a *classification* system to group the millions of different organisms on Earth. **Classification** is putting things into groups based on characteristics the things share. Classification helps scientists answer several important questions:

- What are the defining characteristics of each species?
- When did the characteristics of a species evolve?
- What are the relationships between different species?

STUDY TIP

Organize As you read, make a diagram to show the eight-level system of organization.

Say It

Discuss With a partner, describe some items at home that you have put into groups. Explain why you grouped them and what characteristics you used.

How Do Scientists Classify Organisms?

What are some ways we can classify organisms? Perhaps we could group them by where they live or how they are useful to humans. Throughout history, people have classified organisms in many different ways.

In the 1700s, a Swedish scientist named Carolus Linnaeus created his own system. His system was based on the structure or characteristics of organisms. With his new system, Linnaeus founded modern taxonomy. **Taxonomy** is the science of describing, classifying, and naming organisms. ☑

READING CHECK

1. Explain How did Linnaeus classify organisms?

____

____

____

CLASSIFICATION TODAY

Taxonomists use an eight-level system to classify living things based on shared characteristics. Scientists also use shared characteristics to describe how closely related living things are.

The more characteristics organisms share, the more closely related they may be. For example, the platypus, brown bear, lion, and house cat are thought to be related because they share many characteristics. These animals all have hair and mammary glands, so they are grouped together as mammals. However, they can also be classified into more specific groups.

Critical Thinking

2. Infer What is the main difference between organisms that share many characteristics and organisms that do not?

____

____

____

____

____

____

BRANCHING DIAGRAMS

Shared characteristics can be shown in a *branching diagram*. Each characteristic on the branching diagram is shared by only the animals above it. The characteristics found higher on the diagram evolved more recently than the characteristics below them.

In the diagram below, all of the animals have hair and mammary glands. However, only the brown bear, lion, and house cat give birth to live young. More recent organisms are at the ends of branches high on the diagram. For example, according to the diagram, the house cat evolved more recently than the platypus. ☑

READING CHECK

3. Identify On a branching diagram, where would you see the characteristics that evolved most recently?

____

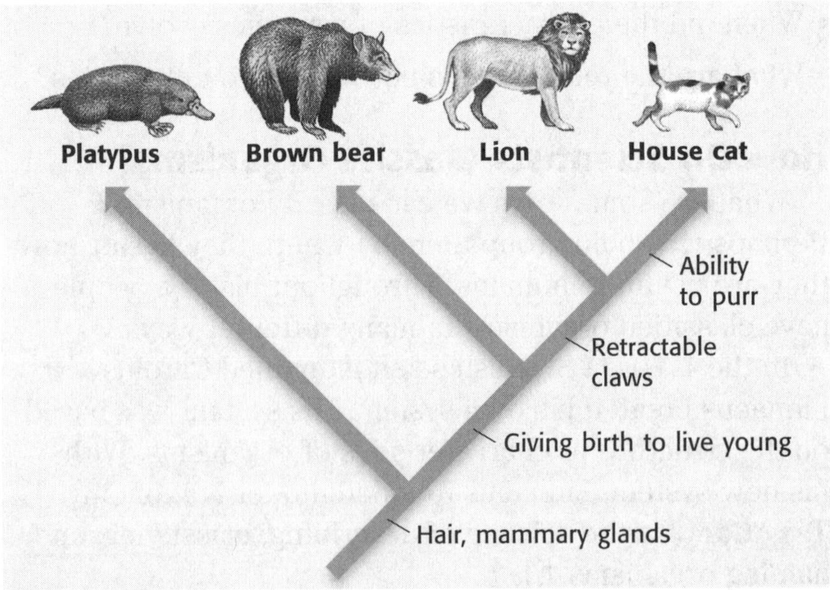
Platypus
Brown bear
Lion
House cat
Ability to purr
Retractable claws
Giving birth to live young
Hair, mammary glands

This branching diagram shows the similarities and differences between four kinds of mammals. The bottom of the diagram begins in the past, and the tips of the branches end in the present.

TAKE A LOOK

4. Identify According to the diagram, which organisms evolved before the lion? Circle these organisms.

What Are the Levels of Classification?

Scientists use shared characteristics to group organisms into eight levels of classification. At each level of classification, there are fewer organisms than in the level above. A domain is the largest, most general level of classification. Every living thing is classified into one of three domains.

Species is the smallest level of classification. A species is a group of organisms that can mate and produce fertile offspring. For example, dogs are all one species. They can mate with one another and have fertile offspring. The figure on the next page shows each of the eight levels of classification.

TWO-PART NAMES

We usually call organisms by common names. For example, "cat," "dog," and "human" are all common names. However, people who speak a language other than English have different names for a cat and dog. Sometimes, organisms are even called by different names in English. For example, cougar, mountain lion, and puma are three names for the same animal! ☑

Scientists need to be sure they are all talking about the same organism. They give organisms *scientific names.* Scientific names are the same in all languages. An organism has only one scientific name.

Scientific names are based on the system created by Linnaeus. He gave each kind of organism a two-part name. The first part of the name is the *genus*, and the second part is the *species*. All genus names begin with a capital letter. All species names begin with a lowercase letter. Both words in a scientific name are underlined or italicized. For example, the scientific name for the Asian elephant is *Elephas maximus.* ☑

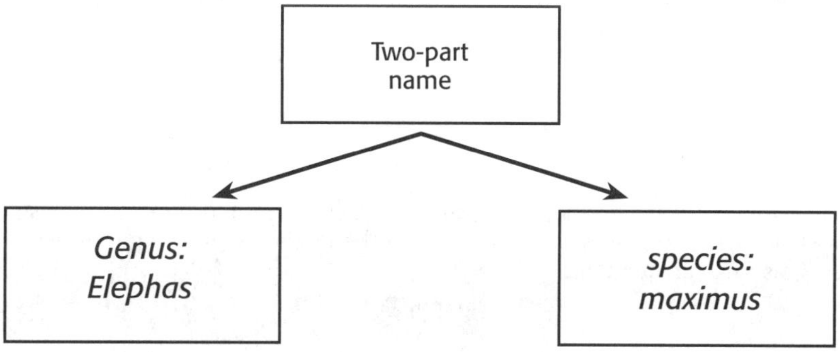
Two-part name
Genus: Elephas
species: maximus

READING CHECK

5. List What are two problems with common names?

____

____

____

____

____

READING CHECK

6. Identify What are the two parts of a scientific name?

____

____

Levels of Classification of the House Cat

Kingdom Animalia: All animals are in the kingdom Animalia.

Phylum Chordata: All animals in the phylum Chordata have a hollow nerve cord. Most have a backbone.

Class Mammalia: Animals in the class Mammalia have a backbone. They also nurse their young.

Order Carnivora: Animals in the order Carnivora have a backbone and nurse their young. They also have special teeth for tearing meat.

Family Felidae: Animals in the family Felidae are cats. They have a backbone, nurse their young, have special teeth for tearing meat, and have retractable claws.

Genus *Felis:* Animals in the genus *Felis* share traits with other animals in the same family. However, these cats cannot roar; they can only purr.

Species *Felis catus*: The species *Felis catus* is the common house cat. The house cat shares traits with all of the organisms in the levels above the species level, but it also has unique traits.

TAKE A LOOK

7. Identify Which level contains organisms that are more closely related: a phylum or a class?

____

8. Describe How does the number of organisms change from the level of kingdom to the level of species?

____

____

____

What Is a Dichotomous Key?

What could you do if you found an organism that you did not recognize? You could use a special guide called a dichotomous key. A **dichotomous key** is set of paired statements that give descriptions of organisms. These statements let you rule out out certain species based on characteristics of your specimen. There are many dichotomous keys for many different kinds of organisms. You could even make your own!

In a dichotomous key, there are only two choices at each step. To use the key, you start with the first pair of statements. You choose the statement from the pair that describes the organism. At each step, the key may identify the organism or it may direct you to another pair of statements. By working through the statements in order, you can identify the organism.

Critical Thinking

9. Infer Why couldn't one single dichotomous key be used for all of the organisms on Earth?

____

____

____

____

____

Dichotomous Key to 10 Common Mammals in the Eastern United States

1. a. This mammal flies. Its "hand" forms a wing. **b.** This mammal does not fly. It's "hand" does not form a wing.	**little brown bat** **Go to step 2.**
2. a. This mammal has no hair on its tail. **b.** This mammal has hair on its tail.	**Go to step 3.** **Go to step 4.**
3. a. This mammal has a short, naked tail. **b.** This mammal has a long, naked tail.	**eastern mole** **Go to step 5.**
4. a. This mammal has a black mask across its face. **b.** This mammal does not have a black mask across its face.	**raccoon** **Go to step 6.**
5. a. This mammal has a tail that is flat and paddle shaped. **b.** This mammal has a tail that is not flat or paddle shaped.	**beaver** **opossum**
6. a. This mammal is brown and has a white underbelly. **b.** This mammal is not brown and does not have a white underbelly.	**Go to step 7.** **Go to step 8.**
7. a. This mammal has a long, furry tail that is black on the tip. **b.** This mammal has a long tail that has little fur.	**longtail weasel** **white-footed mouse**
8. a. This mammal is black and has a narrow white stripe on its forehead and broad white stripes on its back. **b.** This mammal is not black and does not have white stripes.	**striped skunk** **Go to step 9.**
9. a. This mammal has long ears and a short, cottony tail. **b.** This mammal has short ears and a medium-length tail.	**eastern cottontail** **woodchuck**

TAKE A LOOK

10. Identify Use this dichotomous key to identify the two animals shown.

____

____

Name ______________________ Class ____________ Date ____________

Section 1 Review

NSES **LS 5a**

SECTION VOCABULARY

classification the division of organisms into groups, or classes, based on specific characteristics **dichotomous key** an aid that is used to identify organisms and that consists of the answers to a series of questions	**taxonomy** the science of describing, naming, and classifying organisms

1. List Give the eight levels of classification from the largest to the smallest.

____

____

2. Identify According to the branching diagram below, which characteristic do ferns have that mosses do not?

____

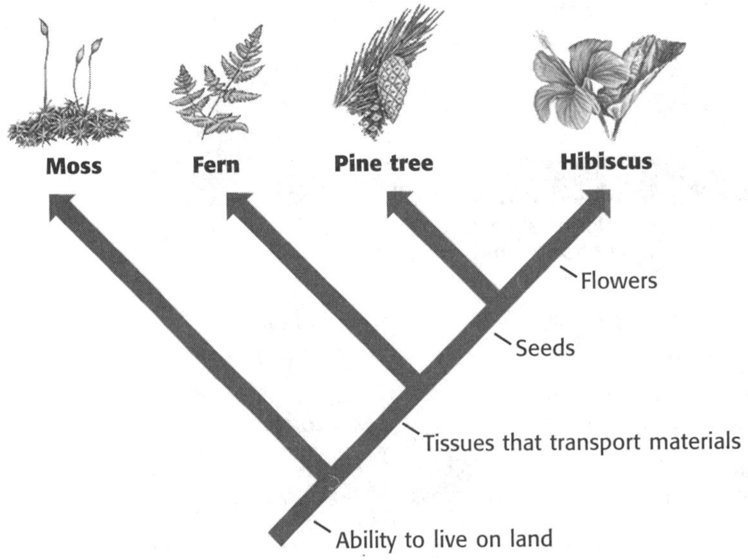

Moss
Fern
Pine tree
Hibiscus
Flowers
Seeds
Tissues that transport materials
Ability to live on land

3. Analyze Which species in the diagram above is most similar to the hibiscus? Which is the least similar?

____

____

4. Identify What are the two parts of a scientific name?

____

5. Infer Could you use the dichotomous key in this section to identify a species of lizard? Explain your answer.

____

____

SECTION 2

Domains and Kingdoms

BEFORE YOU READ

After you read this section, you should be able to answer these questions:

- How are prokaryotes classified?
- How are eukaryotes classified?

National Science Education Standards
LS 1f, 2a, 4b, 4c, 5b

How Do Scientists Classify Organisms?

For hundreds of years, all organisms were classified as either plants or animals. However, as more organisms were discovered, scientists found some organisms that did not fit well into these two kingdoms. Some animals, for example, had characteristics of both plants and animals.

What would you call an organism that is green and makes its own food? Is it a plant? What if the organism moved and could also eat other organisms? Plants generally do neither of these things. Is the organism a plant or an animal?

STUDY TIP

List As you read this section, make a list of the domains and kingdoms scientists use to classify organisms.

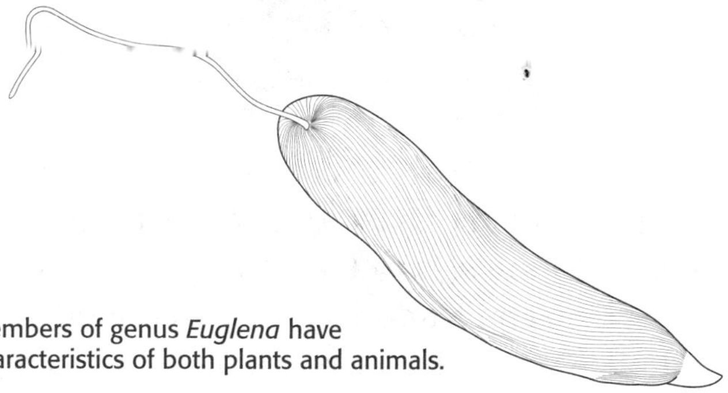

Members of genus *Euglena* have characteristics of both plants and animals.

The organism above belongs to genus *Euglena*. Its members show all of the characteristics just described. As scientists discovered organisms, such as *Euglena*, that didn't fit easily into existing groups, they created new ones. As they added kingdoms, scientists found that members of some kingdoms were closely related to members of other kingdoms. Today, scientists group kingdoms into *domains*.

All organisms on Earth are grouped into three domains. Two domains, Bacteria and Archaea, are made up of prokaryotes. The third domain, Eukarya, is made up of all the eukaryotes. Scientists are still working to describe the kingdoms in each of the three domains.

Critical Thinking

1. Apply Concepts In which domain would multicellular organisms be classified? Explain your answer.

____

____

____

____

How Are Prokaryotes Classified?

A prokaryote is a single-celled organism that does not have a nucleus. Prokaryotes are the oldest group of organisms on Earth. They make up two domains: Archaea and Bacteria.

DOMAIN ARCHAEA

Domain **Archaea** is made up of prokaryotes. The cell walls and cell membranes of archaea are made of different substances than those of other prokaryotes. Many archaea can live in extreme environments where other organisms could not survive. Some archaea can also be found in more moderate environments, such as the ocean. ☑

READING CHECK

2. Compare How are members of Archaea different from other prokaryotes?

____

____

____

____

____

DOMAIN BACTERIA

All bacteria belong to domain **Bacteria**. Bacteria can be found in the air, in soil, in water, and even on and inside the human body!

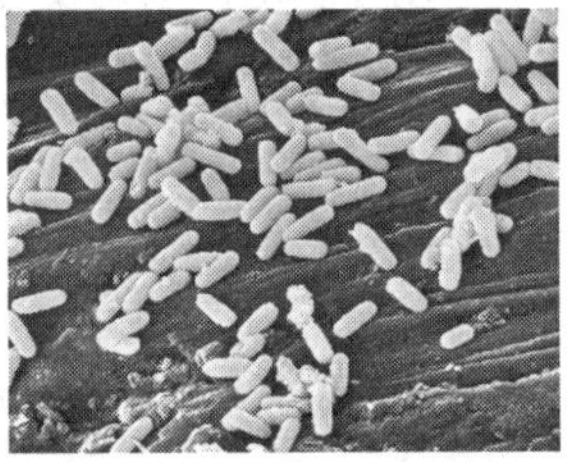

We often think of bacteria as bad, but not all bacteria are harmful. One kind of bacterium changes milk into yogurt. *Escherichia coli* is a bacterium that lives in human intestines. It helps break down undigested food and produces vitamin K. Some bacteria do cause diseases, such as pneumonia. However, other bacteria make chemicals that can help us fight bacteria that cause disease. ☑

READING CHECK

3. Explain Are all bacteria harmful? Explain your answer.

____

____

____

____

The Grand Prismatic Spring in Yellowstone National Park contains water that is about 90°C (190°F). Most organisms would die in such a hot environment.

TAKE A LOOK

4. Apply Concepts What kind of prokaryotes do you think could live in this spring? Explain your answer.

____

____

____

____

How Are Eukaryotes Classified?

Organisms that have cells with membrane-bound organelles and a nucleus are called *eukaryotes*. All eukaryotes belong to domain **Eukarya**. Domain Eukarya includes the following kingdoms: Protista, Fungi, Plantae, and Animalia.

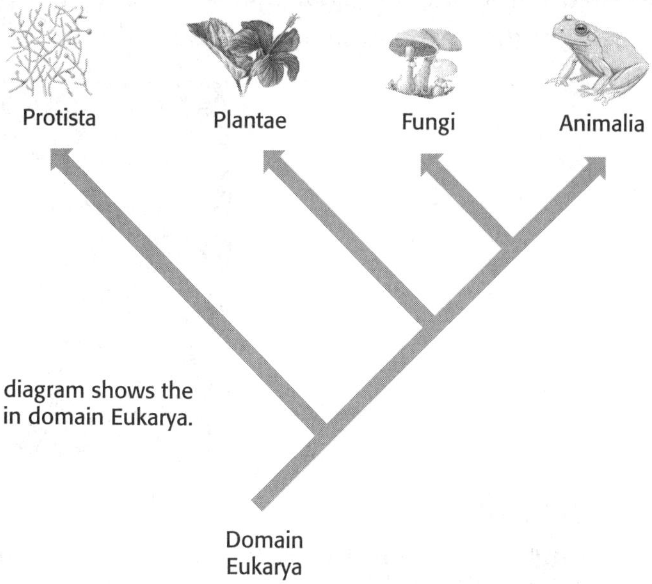
Protista
Plantae
Fungi
Animalia
Domain Eukarya

This branching diagram shows the four kingdoms in domain Eukarya.

STANDARDS CHECK

LS 5a Biological evolution accounts for the diversity of species developed through gradual processes over many generations. Species acquire many of their unique characteristics through biological adaptation, which involves the selection of naturally occurring variations in populations. Biological adaptations include changes in structures, behaviors, or physiology that enhance survival and reproductive success in a particular environment.

Word Help: selection
the process of choosing

Word Help: variation
a difference in the form or function

5. Identify Based on the branching diagram, which two kingdoms in Eukarya evolved most recently? How do you know?

____
____
____
____
____
____

KINGDOM PROTISTA

Members of kingdom **Protista** are either single-celled or simple multicellular organisms. They are commonly called *protists*. Scientists think that the first protists evolved from ancient bacteria about 2 billion years ago. Much later, plants, fungi, and animals evolved from ancient protists.

Kingdom Protista contains many different kinds of organisms. Some, such as *Paramecium*, resemble animals. They are called *protozoa*. Plantlike protists are called *algae*. Some algae, such as phytoplankton, are single cells. Others, such as kelp, are multicellular. Multicellular slime molds also belong to kingdom Protista.

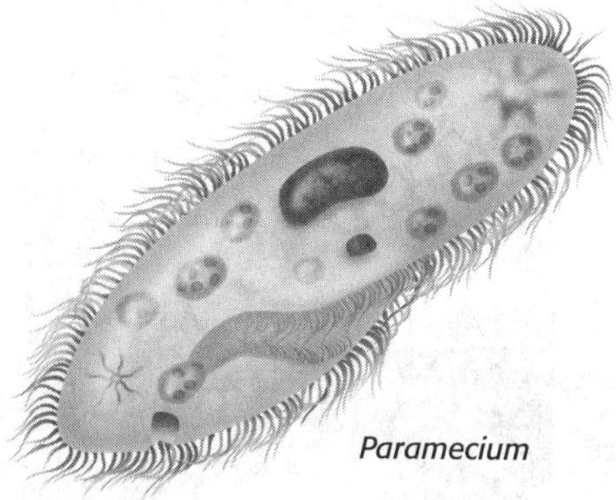

Paramecium

KINGDOM FUNGI

Molds and mushrooms are members of kingdom **Fungi**. Some fungi (singular, *fungus*) are unicellular. That is, they are single-celled organisms. Yeast is unicellular. Most other fungi are multicellular. Unlike plants, fungi do not perform photosynthesis. However, they also do not eat food, as animals do. Instead, fungi use digestive juices to break down materials in the environment and absorb them. ☑

READING CHECK

6. Describe How do fungi get food?

____

____

____

____

Amanita is a poisonous fungus. You should never eat wild fungi.

KINGDOM PLANTAE

Although plants differ in size and appearance, most people can easily identify the members of kingdom Plantae. Kingdom **Plantae** contains organisms that are eukaryotic, have cell walls, and make food by photosynthesis. Most plants need sunlight to carry out photosynthesis. Therefore, plants must live in places where light can reach.

The food that plants make is important for the plants and also for other organisms. Many animals, fungi, protists, and bacteria get nutrients from plants. When they digest the plant material, they get the energy stored by the plant. Plants also provide homes for other organisms.

Math Focus

7. Calculate The average student's arms extend about 1.3 m. How many students would have to join hands to form a human chain around a giant sequoia?

____

The giant sequoia is one of the largest members of kingdom Plantae. A giant sequoia can measure 30 m around the base and grow more than 91 m tall!

KINGDOM ANIMALIA

Kingdom **Animalia** contains complex, multicellular organisms. Organisms in kingdom Animalia are commonly called *animals*. The following are some characteristics of animals:

- Their cells do not have cell walls.
- They are able to move from place to place.
- They have sense organs that help them react quickly to their environment.

TAKE A LOOK

8. Identify Which animal characteristic from above can be seen in this bald eagle?

____

STRANGE ORGANISMS

Some organisms are not easy to classify. For example, some plants can eat other organisms to get nutrition as animals do. Some protists use photosynthesis as plants do but also move around as animals do.

Red Cup Sponge

Critical Thinking

9. Apply Concepts To get nutrients, a Venus' flytrap uses photosynthesis and traps and digests insects. Its cells have cell walls. Into which kingdom would you place this organism? Explain your answer.

____

____

____

____

____

What kind of organism is this red cup sponge? It does not have sense organs and cannot move for most of its life. Because of this, scientists once classified sponges as plants. However, sponges cannot make their own food as plants do. They must eat other organisms to get nutrients. Today, scientists classify sponges as animals. Sponges are usually considered the simplest animals.

Name ______________________ Class ______________ Date ______________

Section 2 Review

NSES **LS 1f, 2a, 4b, 4c, 5b**

SECTION VOCABULARY

Animalia a kingdom made up of complex, multicellular organisms that lack cell walls, can usually move around, and quickly respond to their environment

Archaea in a modern taxonomic system, a domain made up of prokaryotes (most of which are known to live in extreme environments) that are distinguished from other prokaryotes by differences in their genetics and in the makeup of their cell wall; this domain aligns with the traditional kingdom Archaebacteria

Bacteria in a modern taxonomic system, a domain made up of prokaryotes that usually have a cell wall and that usually reproduce by cell division. This domain aligns with the traditional kingdom Eubacteria

Eukarya in a modern taxonomic system, a domain made up of all eukaryotes; this domain aligns with the traditional kingdoms Protista, Fungi, Plantae, and Animalia

Fungi a kingdom made up of nongreen, eukaryotic organisms that have no means of movement, reproduce by using spores, and get food by breaking down substances in their surroundings and absorbing the nutrients

Plantae a kingdom made up of complex, multicellular organisms that are usually green, have cell walls made of cellulose, cannot move around, and use the sun's energy to make sugar by photosynthesis

Protista a kingdom of mostly one-celled eukaryotic organisms that are different from plants, animals, bacteria, and fungi

1. **Compare** What is one major difference between domain Eukarya and domains Bacteria and Archaea?

____

____

2. **Explain** Why do scientists continue to add new kingdoms to their system of classification?

____

____

3. **Analyze Methods** Why do you think Linnaeus did not include classifications for archaea and bacteria?

____

____

____

4. **Apply Concepts** Based on its characteristics described at the beginning of this section, in which kingdom would you classify *Euglena*?

____